Geometry
with CalcChat® and CalcView®

Practice Workbook and Test Prep

- Extra Practice
- Review & Refresh
- Self-Assessments
- Test Prep
- Post-Course Test

Erie, Pennsylvania

Cover Image Dmitriy Rybin/Shutterstock.com

Copyright © by Big Ideas Learning, LLC. All rights reserved.

Permission is hereby granted to teachers to reprint or photocopy in classroom quantities only the pages or sheets in this work that carry a Big Ideas Learning copyright notice, provided each copy made shows the copyright notice. These pages are designed to be reproduced by teachers for use in their classes with accompanying Big Ideas Learning material, provided each copy made shows the copyright notice. Such copies may not be sold and further distribution is expressly prohibited. Except as authorized above, prior written permission must be obtained from Big Ideas Learning, LLC to reproduce or transmit this work or portions thereof in any other form or by any other electronic or mechanical means, including but not limited to photocopying and recording, or by any information storage or retrieval system, unless expressly permitted by copyright law. Address inquiries to Permissions, Big Ideas Learning, LLC, 1762 Norcross Road, Erie, PA 16510.

Big Ideas Learning and *Big Ideas Math* are registered trademarks of Larson Texts, Inc.

Printed in the United States

ISBN 13: 978-1-64727-071-1

6 7 8 9—25 24 23

Contents

About the Practice Workbook and Test Prep............................ viii

Chapter 1 **Basics of Geometry**

1.1 Points, Lines, and Planes... 1

1.2 Measuring and Constructing Segments 3

1.3 Using Midpoint and Distance Formulas 5

1.4 Perimeter and Area in the Coordinate Plane............................... 7

1.5 Measuring and Constructing Angles.. 9

1.6 Describing Pairs of Angles 11

Chapter Self-Assessment 13

Test Prep.. 15

Chapter 2 **Reasoning and Proofs**

2.1 Conditional Statements 19

2.2 Inductive and Deductive Reasoning 21

2.3 Postulates and Diagrams 23

2.4 Algebraic Reasoning 25

2.5 Proving Statements about Segments and Angles....................... 27

2.6 Proving Geometric Relationships 29

Chapter Self-Assessment 31

Test Prep.. 33

Chapter 3 **Parallel and Perpendicular Lines**

3.1 Pairs of Lines and Angles 37

3.2 Parallel Lines and Transversals... 39

3.3 Proofs with Parallel Lines 41

3.4 Proofs with Perpendicular Lines.. 43

3.5 Equations of Parallel and Perpendicular Lines 45

Contents

	Chapter Self-Assessment	47
	Test Prep	49

Chapter 4 Transformations

4.1	Translations	53
4.2	Reflections	55
4.3	Rotations	57
4.4	Congruence and Transformations	59
4.5	Dilations	61
4.6	Similarity and Transformations	63
	Chapter Self-Assessment	65
	Test Prep	67

Chapter 5 Congruent Triangles

5.1	Angles of Triangles	71
5.2	Congruent Polygons	73
5.3	Proving Triangle Congruence by SAS	75
5.4	Equilateral and Isosceles Triangles	77
5.5	Proving Triangle Congruence by SSS	79
5.6	Proving Triangle Congruence by ASA and AAS	81
5.7	Using Congruent Triangles	83
5.8	Coordinate Proofs	85
	Chapter Self-Assessment	87
	Test Prep	89

Chapter 6 Relationships Within Triangles

| 6.1 | Perpendicular and Angle Bisectors | 93 |
| 6.2 | Bisectors of Triangles | 95 |

Contents

6.3	Medians and Altitudes of Triangles	97
6.4	The Triangle Midsegment Theorem	99
6.5	Indirect Proof and Inequalities in One Triangle	101
6.6	Inequalities in Two Triangles	103
	Chapter Self-Assessment	105
	Test Prep	107

Chapter 7 **Quadrilaterals and Other Polygons**

7.1	Angles of Polygons	111
7.2	Properties of Parallelograms	113
7.3	Proving That a Quadrilateral Is a Parallelogram	115
7.4	Properties of Special Parallelograms	117
7.5	Properties of Trapezoids and Kites	119
	Chapter Self-Assessment	121
	Test Prep	123

Chapter 8 **Similarity**

8.1	Similar Polygons	127
8.2	Proving Triangle Similarity by AA	129
8.3	Proving Triangle Similarity by SSS and SAS	131
8.4	Proportionality Theorems	133
	Chapter Self-Assessment	135
	Test Prep	137

Chapter 9 **Right Triangles and Trigonometry**

9.1	The Pythagorean Theorem	141
9.2	Special Right Triangles	143
9.3	Similar Right Triangles	145

Contents

9.4	The Tangent Ratio	147
9.5	The Sine and Cosine Ratios	149
9.6	Solving Right Triangles	151
9.7	Law of Sines and Law of Cosines	153
	Chapter Self-Assessment	155
	Test Prep	157

Chapter 10 **Circles**

10.1	Lines and Segments That Intersect Circles	161
10.2	Finding Arc Measures	163
10.3	Using Chords	165
10.4	Inscribed Angles and Polygons	167
10.5	Angle Relationships in Circles	169
10.6	Segment Relationships in Circles	171
10.7	Circles in the Coordinate Plane	173
10.8	Focus of a Parabola	175
	Chapter Self-Assessment	177
	Test Prep	179

Chapter 11 **Circumference and Area**

11.1	Circumference and Arc Length	183
11.2	Areas of Circles and Sectors	185
11.3	Areas of Polygons	187
11.4	Modeling with Area	189
	Chapter Self-Assessment	191
	Test Prep	193

Contents

Chapter 12 Surface Area and Volume

12.1	Cross Sections of Solids	197
12.2	Volumes of Prisms and Cylinders	199
12.3	Volumes of Pyramids	201
12.4	Surface Areas and Volumes of Cones	203
12.5	Surface Areas and Volumes of Spheres	205
12.6	Modeling with Surface Area and Volume	207
12.7	Solids of Revolution	209
	Chapter Self-Assessment	211
	Test Prep	213

Chapter 13 Probability

13.1	Sample Spaces and Probability	217
13.2	Two-Way Tables and Probability	219
13.3	Conditional Probability	221
13.4	Independent and Dependent Events	223
13.5	Probability of Disjoint and Overlapping Events	225
13.6	Permutations and Combinations	227
13.7	Binomial Distributions	229
	Chapter Self-Assessment	231
	Test Prep	233
	Post-Course Test	237

About the Practice Workbook and Test Prep

Extra Practice

The Extra Practice exercises provide additional practice on the key concepts taught in each section.

Review & Refresh

The Review & Refresh exercises provide students the opportunity to practice prior skills to improve retention.

Self-Assessments

For every section and chapter, students can rate their understanding of the learning targets and success criteria.

Test Prep

Each chapter contains a cumulative test to prepare students for standardized test questions, including multiple choice, multi-select, gridded response, and fill-in-the-blank.

Post-Course Test

The Post-Course Test measures students' understanding of all content in this course. This assessment is designed to prepare students for standardized test questions, including multiple choice, multi-select, gridded response, and fill-in-the-blank.

Name_____ Date_____

1.1 Extra Practice

In Exercises 1–4, use the diagram.

1. Give two other names for $\overrightarrow{CD}$.

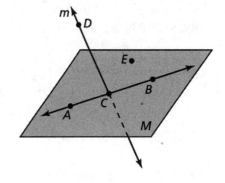

2. Give another name for plane M.

3. Name three points that are collinear. Then name a fourth point that is not collinear with these three points.

4. Name a point that is not coplanar with points A, C, and E.

In Exercises 5–8, use the diagram.

5. What is another name for $\overleftrightarrow{PQ}$?

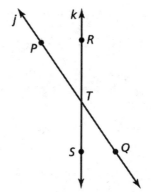

6. What is another name for $\overrightarrow{RS}$?

7. Name all rays with endpoint T. Which of these rays are opposite rays?

8. In the diagram, draw planes M and N that intersect in line k.

In Exercises 9 and 10, sketch the figure described.

9. $\overrightarrow{AB}$ and $\overrightarrow{BC}$

10. line k in plane M

Name _____ Date _____

1.1 Review & Refresh

1. Determine which of the lines, if any, are parallel or perpendicular. Explain.

   Line *a* passes through (–2, 0) and (1, 6).
   Line *b* passes through (–3, 5) and (3, 1).
   Line *c* passes through (1, 1) and (4, 7).

2. Solve $4 + x = 15$.

3. A robot vacuum cleans at a constant speed of 4 feet per second. It will travel 600 feet, plus or minus 40 feet. Write and solve an equation to find the minimum and maximum numbers of seconds it cleans.

4. Evaluate $\sqrt[3]{27^2}$.

5. Graph $f(x) = \frac{1}{2}x - 4$ and $g(x) = f(x + 2)$. Describe the transformation from the graph of *f* to the graph of *g*.

In Exercises 6–9, use the diagram.

6. Name two lines.

7. Name three collinear points.

8. Give two names for the plane.

9. Name two line segments.

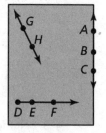

10. Use zeros to graph $y = x(x - 1)(x + 3)$.

11. Make a box-and-whisker plot that represents the data.

    Minutes at an event: 40, 45, 55, 50, 65, 60, 50, 45, 60, 50

1.1 Self-Assessment

Use the scale to rate your understanding of the learning target and the success criteria.

| 1 I do not understand. | 2 I can do it with help. | 3 I can do it on my own. | 4 I can teach someone else. |

	Rating	Date
1.1 Points, Lines, and Planes		
Learning Target: Use defined terms and undefined terms.	1 2 3 4	
I can describe a point, a line, and a plane.	1 2 3 4	
I can define and name segments and rays.	1 2 3 4	
I can sketch intersections of lines and planes.	1 2 3 4	

1.2 Extra Practice

1. Use a ruler to measure the length of the segment to the nearest tenth of a centimeter.

In Exercises 2–4, plot the points in the coordinate plane. Then determine whether $\overline{AB}$ and $\overline{CD}$ are congruent.

2. $A(-5, 5)$, $B(-2, 5)$,
   $C(2, -4)$, $D(-1, -4)$

3. $A(4, 0)$, $B(4, 3)$,
   $C(-4, -4)$, $D(-4, 1)$

4. $A(-1, 5)$, $B(5, 5)$,
   $C(1, 3)$, $D(1, -3)$

In Exercises 5–7, find VW.

5.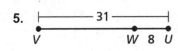
6.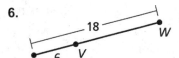
7.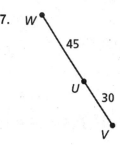

8. A bookstore and a movie theater are 6 kilometers apart along the same street. A florist is located between the bookstore and the theater on the same street. The florist is 2.5 kilometers from the theater. How far is the florist from the bookstore?

Name _____ Date _____

1.2 Review & Refresh

In Exercises 1 and 2, solve the equation.

1. $-4 + y = 1$
2. $-7x = 28$

3. Write an inequality that represents the graph.

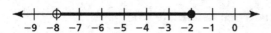

4. Use intercepts to graph the linear equation $-2x + 4y = 16$. Label the points corresponding to the intercepts.

5. Determine whether the relation is a function. Explain.

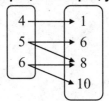

6. Graph $f(x) = 4 - 2^x$. Identify the asymptote. Find the domain and range of f.

7. Find BC.

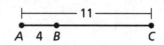

8. A photographer takes portraits of seniors and young professionals. The photographer takes 8 portraits earning a total of $310 in one week. Each senior portrait is worth $20 and each young professional portrait is worth $50. How many senior portraits does the photographer take? young professional portraits?

9. Write an equation in slope-intercept form of the line that passes through $(0, 7)$ and $(2, 3)$.

1.2 Self-Assessment

Use the scale to rate your understanding of the learning target and the success criteria.

| 1 I do not understand. | 2 I can do it with help. | 3 I can do it on my own. | 4 I can teach someone else. |

	Rating	Date
1.2 Measuring and Constructing Segments		
Learning Target: Measure and construct line segments.	1 2 3 4	
I can measure a line segment.	1 2 3 4	
I can copy a line segment.	1 2 3 4	
I can explain and use the Segment Addition Postulate.	1 2 3 4	

Name_____ Date_____

1.3 Extra Practice

In Exercises 1–3, identify the segment bisector of $\overline{AB}$. Then find AB.

1.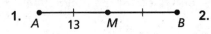
2.
3.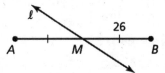

In Exercises 4–6, identify the segment bisector of $\overline{EF}$. Then find EF.

4.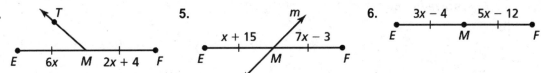
5.
6.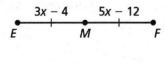

In Exercises 7–9, the endpoints of $\overline{PQ}$ are given. Find the coordinates of the midpoint M.

7. $P(-4, 3)$ and $Q(0, 5)$
8. $P(-2, 7)$ and $Q(10, -3)$
9. $P(3, -15)$ and $Q(9, -3)$

In Exercises 10–12, the midpoint M and one endpoint of $\overline{JK}$ are given. Find the coordinates of the other endpoint.

10. $J(7, 2)$ and $M(1, -2)$
11. $M(0, -1)$ and $K(-5, 0)$
12. $J(2, 16)$ and $M\left(-\frac{9}{2}, 7\right)$

13. A botanical garden is 3 miles west and 1 mile north of your apartment. A metro station is 1 mile east and 1 mile south of your apartment. Estimate the distance between the botanical garden and the metro station.

Name _____ Date _____

1.3 Review & Refresh

In Exercises 1 and 2, find the perimeter and area of the figure.

1.
2.

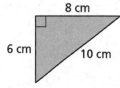

In Exercises 3 and 4, solve the inequality. Graph the solution.

3. $a + 3 < 7$

4. $\dfrac{z}{6} \geq 1$

5. The endpoints of $\overline{CD}$ are $C(-2, 4)$ and $D(4, -4)$. Find the coordinates of the midpoint M. Then find CD.

6. Solve the literal equation $18x + 3y = 6$ for y.

In Exercises 7 and 8, factor the polynomial.

7. $7z^2 - 21z$

8. $81x^2 - 25$

9. Name two pairs of opposite rays in the diagram.

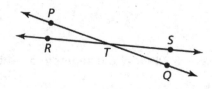

10. Simplify $\dfrac{d^{-3} \cdot d^{11}}{d^7}$. Write your answer using only positive exponents.

11. The function $p(x) = 50 - 2x$ represents the number of party favors remaining after x guests are served. How many favors remain after 5 guests are served?

12. Convert 3.8 kilometers to meters.

1.3 Self-Assessment

Use the scale to rate your understanding of the learning target and the success criteria.

| 1 I do not understand. | 2 I can do it with help. | 3 I can do it on my own. | 4 I can teach someone else. |

	Rating	Date
1.3 Using Midpoint and Distance Formulas		
Learning Target: Find midpoints and lengths of segments.	1 2 3 4	
I can find lengths of segments.	1 2 3 4	
I can construct a segment bisector.	1 2 3 4	
I can find the midpoint of a segment.	1 2 3 4	

Name_____ Date_____

1.4 Extra Practice

In Exercises 1–4, classify the polygon by the number of sides. Tell whether it is *convex* or *concave*.

1.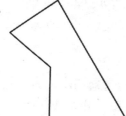
2.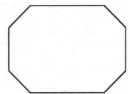
3.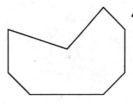
4.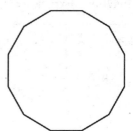

In Exercises 5 and 6, find the perimeter and area of the polygon with the given vertices.

5. $X(2, 4)$, $Y(0, -2)$, $Z(2, -2)$

6. $P(1, 3)$, $Q(1, 1)$, $R(-4, 2)$

7. Find the area of rectangle *JKLM* with vertices $J(-4, 1)$, $K(-4, -2)$, $L(6, -2)$, and $M(6, 1)$.

8. The diagram shows the vertices of a garden. Each unit in the coordinate plane represents 10 feet. Find the value of *a* so that the perimeter of the garden is 300 feet.

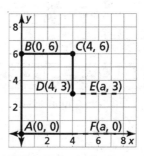

Name _____ Date _____

1.4 Review & Refresh

1. Does the table represent a *linear* or *nonlinear* function? Explain.

x	−2	−1	0	1	2
y	1	4	7	10	13

In Exercises 2 and 3, solve the equation.

2. $x + 2 = 2x - 8$ 3. $\dfrac{x+1}{2} = -3$

4. Give another name for $\overline{CD}$.

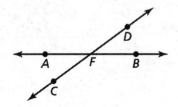

5. The endpoints of a segment are $J(-2, 3)$ and $K(-4, -1)$. Find the coordinates of the midpoint M and the length of the segment.

6. You deposit $100 into a savings account that earns 3% annual interest compounded monthly. Write a function that represents the balance y (in dollars) after t years.

7. Graph $g(x) = |x + 2| - 1$. Then describe the transformations from the graph of $f(x) = |x|$ to the graph of g.

8. Find the perimeter and area of rectangle $ABCD$ with vertices $A(1, 3)$, $B(1, 5)$, $C(6, 3)$, and $D(6, 5)$.

1.4 Self-Assessment

Use the scale to rate your understanding of the learning target and the success criteria.

| 1 | I do not understand. | 2 | I can do it with help. | 3 | I can do it on my own. | 4 | I can teach someone else. |

	Rating	Date
1.4 Perimeter and Area in the Coordinate Plane		
Learning Target: Find perimeters and areas of polygons in the coordinate plane.	1 2 3 4	
I can classify and describe polygons.	1 2 3 4	
I can find perimeters of polygons in the coordinate plane.	1 2 3 4	
I can find areas of polygons in the coordinate plane.	1 2 3 4	

Name_____ Date _____

1.5 Extra Practice

1. Name three different angles in the diagram.

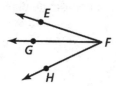

2. Find each angle measure. Then classify the angle.

   a. $m\angle ABD$

   b. $m\angle EBD$

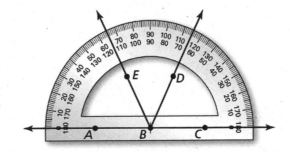

In Exercises 3–8, find the indicated angle measure(s).

3. Find $m\angle JKL$.

4. $m\angle RSU = 91°$. Find $m\angle RST$.

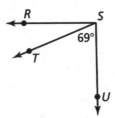

5. $\angle UWX$ is a straight angle. Find $m\angle UWV$ and $m\angle XWV$.

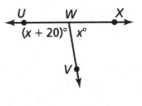

6. Find $m\angle CAD$ and $m\angle BAD$.

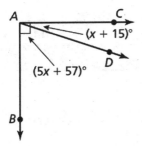

7. $\overrightarrow{EG}$ bisects $\angle DEF$. Find $m\angle DEG$ and $m\angle GEF$.

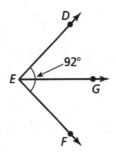

8. $\overrightarrow{QR}$ bisects $\angle PQS$. Find $m\angle PQR$ and $m\angle PQS$.

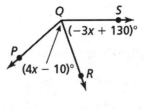

Name _____ Date _____

1.5 Review & Refresh

1. Find the perimeter and the area of $\triangle ABC$ with vertices $A(-2, 2)$, $B(-2, -5)$, and $C(3, 2)$.

2. Solve $2(x - 3) + 1 = 5$.

3. Simplify $\sqrt{18}$.

4. The positions of three pillars are shown. Cords connect Pillar A to Pillar B and Pillar B to Pillar C. Which cord is longer? About how far is Pillar A from Pillar C?

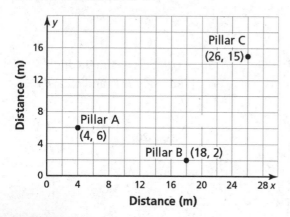

5. Graph $y \geq -x + 1$ in a coordinate plane.

6. $\overrightarrow{KM}$ bisects $\angle JKL$. Find $m\angle JKL$.

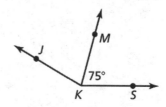

7. Solve the system using any method. Explain your choice of method.

   $y = 2x + 1$
   $-2y + x = 7$

8. Point Y is between points X and Z on $\overline{XZ}$. $XY = 14$ and $YZ = 18$. Find XZ.

1.5 Self-Assessment

Use the scale to rate your understanding of the learning target and the success criteria.

| 1 I do not understand. | 2 I can do it with help. | 3 I can do it on my own. | 4 I can teach someone else. |

	Rating	Date
1.5 Measuring and Constructing Angles		
Learning Target: Measure, construct, and describe angles.	1 2 3 4	
I can measure and classify angles.	1 2 3 4	
I can construct congruent angles.	1 2 3 4	
I can find angle measures.	1 2 3 4	
I can construct an angle bisector.	1 2 3 4	

Name_____ Date_____

1.6 Extra Practice

In Exercises 1 and 2, use the diagrams.

1. Name the pair(s) of adjacent complementary angles.

2. Name the pair(s) of nonadjacent supplementary angles.

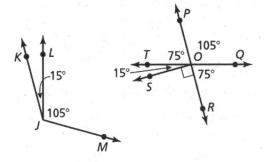

In Exercises 3 and 4, find the angle measure.

3. ∠A is a complement of ∠B, and m∠A = 36°. Find m∠B.

4. ∠C is a supplement of ∠D, and m∠D = 117°. Find m∠C.

In Exercises 5 and 6, find the measure of each angle.

5.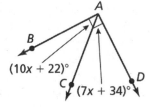

6.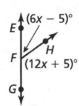

In Exercises 7–9, use the diagram.

7. Identify all the linear pairs that include ∠1.

8. Identify all the vertical angles.

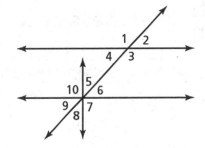

9. Are ∠6 and ∠7 a linear pair? Explain.

10. Two angles form a linear pair. The measure of one angle is three times the measure of the other angle. Find the measure of each angle.

1.6 Review & Refresh

1. Find the area of □ABCD with vertices A(1, 2), B(3, 6), C(7, 6), and D(5, 2).

2. The midpoint of $\overline{JK}$ is M(1, 3). One endpoint is K(5, 4). Find the coordinates of endpoint J.

3. Identify the segment bisector of $\overline{DF}$. Then find DF.

4. Solve $|z - 4| = 1$. Graph the solution.

5. Find the product of $(x + 2)$ and $(3x - 1)$.

6. The total cost (in dollars) of renting a karaoke room for x hours is represented by the function $f(x) = 10x + 10$. The hourly rate is increased by 50%. The new total cost is represented by the function $g(x) = f\left(\frac{3}{2}x\right)$. Describe the transformation from the graph of f to the graph of g.

7. Find the slope and the y-intercept of the graph of $y + 2 = -3x$.

8. Given that $m\angle EFG = 62°$, find $m\angle EFH$ and $m\angle HFG$.

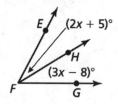

9. ∠1 is a complement of ∠2, and $m\angle 2 = 42°$. Find $m\angle 1$.

1.6 Self-Assessment

Use the scale to rate your understanding of the learning target and the success criteria.

| 1 I do not understand. | 2 I can do it with help. | 3 I can do it on my own. | 4 I can teach someone else. |

	Rating	Date
1.6 Describing Pairs of Angles		
Learning Target: Identify and use pairs of angles.	1 2 3 4	
I can identify complementary and supplementary angles.	1 2 3 4	
I can identify linear pairs and vertical angles.	1 2 3 4	
I can find angle measures in pairs of angles.	1 2 3 4	

Name_____ Date_____

 Chapter Self-Assessment

Use the scale to rate your understanding of the learning target and the success criteria.

| 1 | I do not understand. | 2 | I can do it with help. | 3 | I can do it on my own. | 4 | I can teach someone else. |

	Rating	Date
Chapter 1 Basics of Geometry		
Learning Target: Understand basics of geometry.	1 2 3 4	
I can name points, lines, and planes.	1 2 3 4	
I can measure segments and angles.	1 2 3 4	
I can use formulas in the coordinate plane.	1 2 3 4	
I can construct segments and angles.	1 2 3 4	
1.1 Points, Lines, and Planes		
Learning Target: Use defined terms and undefined terms.	1 2 3 4	
I can describe a point, a line, and a plane.	1 2 3 4	
I can define and name segments and rays.	1 2 3 4	
I can sketch intersections of lines and planes.	1 2 3 4	
1.2 Measuring and Constructing Segments		
Learning Target: Measure and construct line segments.	1 2 3 4	
I can measure a line segment.	1 2 3 4	
I can copy a line segment.	1 2 3 4	
I can explain and use the Segment Addition Postulate.	1 2 3 4	
1.3 Using Midpoint and Distance Formulas		
Learning Target: Find midpoints and lengths of segments.	1 2 3 4	
I can find lengths of segments.	1 2 3 4	
I can construct a segment bisector.	1 2 3 4	
I can find the midpoint of a segment.	1 2 3 4	

Name _____ Date _____

 Chapter Self-Assessment (continued)

	Rating	Date
1.4 Perimeter and Area in the Coordinate Plane		
Learning Target: Find perimeters and areas of polygons in the coordinate plane.	1 2 3 4	
I can classify and describe polygons.	1 2 3 4	
I can find perimeters of polygons in the coordinate plane.	1 2 3 4	
I can find areas of polygons in the coordinate plane.	1 2 3 4	
1.5 Measuring and Constructing Angles		
Learning Target: Measure, construct, and describe angles.	1 2 3 4	
I can measure and classify angles.	1 2 3 4	
I can construct congruent angles.	1 2 3 4	
I can find angle measures.	1 2 3 4	
I can construct an angle bisector.	1 2 3 4	
1.6 Describing Pairs of Angles		
Learning Target: Identify and use pairs of angles.	1 2 3 4	
I can identify complementary and supplementary angles.	1 2 3 4	
I can identify linear pairs and vertical angles.	1 2 3 4	
I can find angle measures in pairs of angles.	1 2 3 4	

Chapter 1 Test Prep

1. Select all the expressions that are greater than $|-2|$.

   Ⓐ 0

   Ⓑ $|-2.1|$

   Ⓒ $|-0.5 \cdot 4|$

   Ⓓ $|-1.9|$

   Ⓔ $-|4|+|-8|$

2. What is the common ratio of the sequence 2, −4, 8, −16, 32, …?

   Ⓐ −6

   Ⓑ −2

   Ⓒ $-\dfrac{1}{2}$

   Ⓓ 2

3. For which set of points is $\overline{AB} \cong \overline{CD}$?

   Ⓐ $A(-2, 1)$, $B(-2, 5)$, $C(1, -4)$, $D(1, 3)$

   Ⓑ $A(0, 3)$, $B(4, 2)$, $C(0, 5)$, $D(4, 3)$

   Ⓒ $A(0, 2)$, $B(0, 1)$, $C(4, 1)$, $D(5, 2)$

   Ⓓ $A(-1, -4)$, $B(4, -4)$, $C(-3, -6)$, $D(-3, -1)$

4. Evaluate $h(x) = 6.1 - 3.4x$ when $x = -2$.

   $h(-2) =$

5. Find the distance from $(0, 0)$ to $(3, 4)$.

   _____ units

Chapter 1 Test Prep (continued)

6. Find the area of the polygon.
   - Ⓐ 9 square units
   - Ⓑ 12 square units
   - Ⓒ 17 square units
   - Ⓓ 24 square units

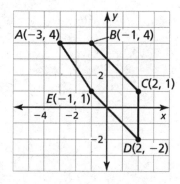

7. Which square root is in simplest form?
   - Ⓐ $\sqrt{9}$
   - Ⓑ $\sqrt{8}$
   - Ⓒ $\sqrt{30}$
   - Ⓓ $\sqrt{48}$

8. What is the solution of $|x - 5| \leq 2$?
   - Ⓐ $3 \leq x \leq 7$
   - Ⓑ $-3 \leq x \leq 7$
   - Ⓒ $x \leq 3$ or $x \geq 7$
   - Ⓓ $x \leq -7$ or $x \geq 3$

9. What is the solution of the system $2x - y = 4$ and $3x - 2y = 2$?
   - Ⓐ $(3, 2)$
   - Ⓑ $(6, 8)$
   - Ⓒ $(10, 14)$
   - Ⓓ $(0, -1)$

10. The ratio of the measure of an angle to the measure of its complement is 2 : 3. What are the two angle measures?
    - Ⓐ 30° and 45°
    - Ⓑ 30° and 60°
    - Ⓒ 36° and 54°
    - Ⓓ 72° and 108°

11. An object is launched from the top of a hill and travels in a parabolic path until it reaches the ground. Write a quadratic function that models the path of the object with a maximum height of 200 feet, represented by a vertex of $(25, 200)$, landing at the point $(55, 0)$.

Chapter 1 Test Prep (continued)

12. Which equation is written in point-slope form?

    Ⓐ $y - 2 = 3(x - 3)$

    Ⓑ $y + 1 = 2x + 5$

    Ⓒ $y = -3x + 4$

    Ⓓ $y = -7$

13. The endpoints of $\overline{AB}$ are $A(2, 5)$ and $B(-4, 7)$. Find the coordinates of the midpoint.

    Ⓐ $(3, -1)$

    Ⓑ $(-2, 12)$

    Ⓒ $(-1, 6)$

    Ⓓ $(3, 6)$

14. Select all the points that are collinear with points Q and T.

    Ⓐ X

    Ⓑ S

    Ⓒ R

    Ⓓ W

    Ⓔ V

15. What is the domain of the function $y = 5\sqrt{6 - 2x}$?

    Ⓐ $x \leq 3$

    Ⓑ $x < 3$

    Ⓒ $x > 3$

    Ⓓ $x \geq 3$

16. Which data set has the least median?

    Ⓐ 0, 2, 3, 3, 0, 1

    Ⓑ 1, 2, 5, 5, 6, 0, 0

    Ⓒ 0, 0, 5, 6, 5, 0, 5

    Ⓓ 7, 0, 9, 9, 1, 1

17. Find the measure of $\angle LNK$. Then classify the angle as *acute*, *right*, or *obtuse*.

    _____

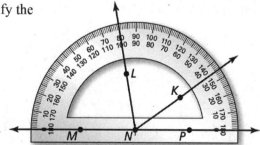

Chapter 1 Test Prep (continued)

18. ∠1 and ∠5 are vertical angles, ∠5 and ∠6 are supplementary angles, and $m\angle 6 = 110.2°$. What is $m\angle 1$?

degrees

19. The electric current I (in amperes) an appliance uses is given by $I = \sqrt{\dfrac{P}{R}}$, where P is the power (in watts) and R is the resistance (in ohms). Find P when $I = 5$ amperes and $R = 4$ ohms.

watts

20. A container in the shape of a rectangular prism has a volume of 1080 cubic inches. What are the dimensions of the container?

Ⓐ $\ell = 18$ in., $w = 180$ in., $h = 188$ in.

Ⓑ $\ell = 18$ in., $w = 12.8$ in., $h = 16.8$ in.

Ⓒ $\ell = 18$ in., $w = 6$ in., $h = 10$ in.

Ⓓ $\ell = 18$ in., $w = 10$ in., $h = 14$ in.

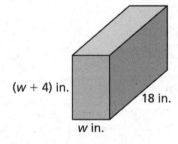

21. ∠BCA and ∠ACD form a linear pair. ∠BCA is represented by $(2x - 5)°$ and ∠ACD is represented by $(3x + 10)°$. What is $m\angle ACD$?

Ⓐ 35°

Ⓑ 61°

Ⓒ 65°

Ⓓ 115°

18 Geometry
Practice Workbook and Test Prep

2.1 Extra Practice

In Exercises 1 and 2, rewrite the conditional statement in if-then form.

1. $13x - 5 = -18$, because $x = -1$.

2. The sum of the measures of the interior angles of a triangle is 180°.

3. Let p be "quadrilateral $ABCD$ is a rectangle" and let q be "the sum of the angle measures is 360°." Write the conditional statement $p \to q$, the converse $q \to p$, the inverse $\sim p \to \sim q$, and the contrapositive $\sim q \to \sim p$ in words. Then decide whether each statement is *true* or *false*.

In Exercises 4–6, decide whether the statement about the diagram is true. Explain your answer using the definitions you have learned.

4. $\overline{LM}$ bisects $\overline{JK}$.

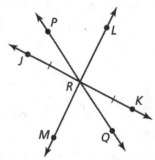

5. $\angle JRP$ and $\angle PRL$ are complementary.

6. $\angle MRQ \cong \angle PRL$

7. An angle is acute when its measure is greater than 0° and less than 90°. Rewrite this definition as a biconditional statement.

2.1 Review & Refresh

1. Write the next three terms of the arithmetic sequence 2, 5, 8, 11, ….

2. Determine whether the graph represents a function. Explain.

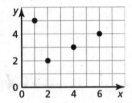

3. Use the graphs of f and g to describe the transformation from the graph of $f(x) = -x + 2$ to the graph of $g(x) = 2f(x)$.

4. Two angles form a linear pair. The measure of one angle is nineteen times the measure of the other angle. Find the measure of each angle.

5. Find the perimeter and the area of $\triangle ABC$ with vertices $A(1, 2)$, $B(1, 6)$, and $C(3, 4)$.

6. The distance from one asteroid to another is 1.876×10^6 kilometers. Write this number in standard form.

7. In the diagram, $\overrightarrow{DB}$ bisects $\angle ADC$, and $m\angle ADB = 56°$. Find $m\angle CDB$ and $m\angle ADC$.

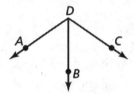

8. Find $(3m^2 - 8) - (2m^2 + 5m)$.

9. Write an inequality that represents the graph.

10. Let p be "you post a video" and let q be "your video goes viral." Write the conditional statement $p \rightarrow q$.

2.1 Self-Assessment

Use the scale to rate your understanding of the learning target and the success criteria.

| 1 I do not understand. | 2 I can do it with help. | 3 I can do it on my own. | 4 I can teach someone else. |

	Rating	Date
2.1 Conditional Statements		
Learning Target: Understand and write conditional statements.	1 2 3 4	
I can write conditional statements.	1 2 3 4	
I can write biconditional statements.	1 2 3 4	
I can determine if conditional statements are true by using truth tables.	1 2 3 4	

2.2 Extra Practice

In Exercises 1 and 2, describe the pattern. Then write or draw the next two numbers or figures.

1. 20, 19, 17, 14, 10, …

2.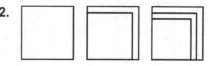

In Exercises 3 and 4, make and test a conjecture about the given quantity.

3. the sum of two negative integers

4. the product of three consecutive nonzero integers

In Exercises 5 and 6, find a counterexample to show that the conjecture is false.

5. If n is a rational number, then n^2 is always less than n.

6. Line k intersects plane P at point Q on the plane. Plane P is perpendicular to line k.

In Exercises 7 and 8, use the Law of Detachment to determine what you can conclude from the given information, if possible.

7. If a triangle has equal side lengths, then each interior angle measure is $60°$. $\triangle ABC$ has equal side lengths.

8. If a quadrilateral is a rhombus, then it has two pairs of opposite sides that are parallel. Quadrilateral $PQRS$ has two pairs of opposite sides that are parallel.

In Exercises 9 and 10, use the Law of Syllogism to write a new conditional statement that follows from the pair of true statements, if possible.

9. If it does not rain, then I will walk to school.

   If I walk to school, then I will wear my sneakers.

10. If $x > 1$, then $3x > 3$.

    If $3x > 3$, then $(3x)^2 > 3^2$.

2.2 Review & Refresh

1. Identify the hypothesis and the conclusion. Then rewrite the conditional statement in if-then form.

   Fracking causes microearthquakes.

2. Classify the polygon by the number of sides. Tell whether it is *convex* or *concave*.

3. Write a recursive rule for the sequence.

n	1	2	3	4
a_n	25	21	17	13

4. Write an equation of the line that passes through the points $(2, 2)$ and $(0, 4)$.

5. Determine whether the equation $y = 2x^2$ represents a *linear* or *nonlinear* function.

6. Solve the equation $3x - 4 = -2x + 6$.

7. Graph $g(x) = 4x^2$. Compare the graph to the graph of $f(x) = x^2$.

8. Approximate $\sqrt{21}$ to the nearest (a) integer and (b) tenth.

9. Find $(3p^2 + p - 1) + (2p^2 - p - 3)$.

10. Solve $7^{2x+3} = 7^{13}$.

2.2 Self-Assessment

Use the scale to rate your understanding of the learning target and the success criteria.

| 1 | I do not understand. | 2 | I can do it with help. | 3 | I can do it on my own. | 4 | I can teach someone else. |

2.2 Inductive and Deductive Reasoning	Rating	Date
Learning Target: Use inductive and deductive reasoning.	1 2 3 4	
I can use inductive reasoning to make conjectures.	1 2 3 4	
I can use deductive reasoning to verify conjectures.	1 2 3 4	
I can distinguish between inductive and deductive reasoning.	1 2 3 4	

Name_____ Date_____

2.3 Extra Practice

In Exercises 1 and 2, state the postulate illustrated by the diagram.

1.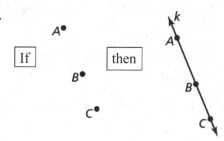

2.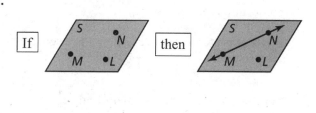

In Exercises 3 and 4, use the diagram to write an example of the postulate.

3. Plane-Point Postulate

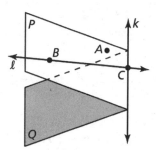

4. Plane Intersection Postulate

In Exercises 5 and 6, sketch a diagram of the description.

5. $\overrightarrow{RS}$ bisecting $\overline{KL}$ at point R

6. $\overleftrightarrow{AB}$ in plane U intersecting $\overleftrightarrow{CD}$ at point E, and point C not in plane U

In Exercises 7–10, use the diagram to determine whether you can assume the statement.

7. Planes A and B intersect at $\overrightarrow{EF}$.

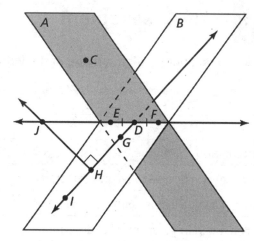

8. $\overleftrightarrow{HJ}$ and $\overleftrightarrow{ID}$ are perpendicular.

9. $\overrightarrow{GD}$ is a bisector of $\overleftrightarrow{EF}$ at point D.

10. $\overline{IH} \cong \overline{HG}$

Name_____ Date _____

2.3 Review & Refresh

1. Find a counterexample to show that the conjecture is false.

   If a quadrilateral is convex, then it is a square.

In Exercises 2 and 3, solve the equation. Justify each step.

2. $z + 1 = 2$
3. $-2 = \dfrac{y}{3}$

4. $\angle 1$ is a supplement of $\angle 2$, and $m\angle 2 = 87°$. Find $m\angle 1$.

5. A portable charger in the shape of a rectangular prism has a width of 3 inches. Its length is twice its height. The volume of the charger is 24 cubic inches. Find the length and height of the portable charger.

6. Write the next three terms of the geometric sequence $-\dfrac{1}{3}, -1, -3, -9, \ldots$.

7. Rewrite the statements as a single biconditional statement.

   If your phone vibrates, then you have an unread notification.

   If you have an unread notification, then your phone vibrates.

In Exercises 8–10, use the diagram to determine whether you can assume the statement.

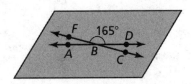

8. Points A, B, and C are coplanar.

9. $m\angle CBD = 15°$

10. $\overrightarrow{FB}$ bisects $\overline{AD}$.

2.3 Self-Assessment

Use the scale to rate your understanding of the learning target and the success criteria.

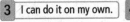

1 I do not understand.   2 I can do it with help.   3 I can do it on my own.   4 I can teach someone else.

	Rating	Date
2.3 Postulates and Diagrams		
Learning Target: Interpret and sketch diagrams.	1 2 3 4	
I can identify postulates represented by diagrams.	1 2 3 4	
I can sketch a diagram given a verbal description.	1 2 3 4	
I can interpret a diagram.	1 2 3 4	

2.4 Extra Practice

In Exercises 1–3, solve the equation. Justify each step.

1. $3x - 7 = 5x - 19$

2. $20 - 2(3x - 1) = x - 6$

3. $-5(2u + 10) = 2(u - 7)$

In Exercises 4 and 5, solve the equation for the given variable. Justify each step.

4. $9x + 2y = 5$; y

5. $\frac{1}{15}s - \frac{2}{3}t = -2$; s

6. The formula for the surface area S of a cone is $S = \pi r^2 + \pi r \ell$, where r is the radius and ℓ is the slant height. Solve the formula for ℓ. Justify each step. Then find the slant height of a cone with a surface area of 113 square feet and a radius of 4 feet. Approximate to the nearest tenth.

7. A semicircular desk at a customer service office is separated into four work regions as shown, such that $m\angle AFD = m\angle EFB$. Show that the left desk region is equal to the right desk region.

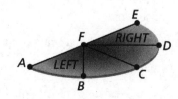

2.4 Review & Refresh

1. Name the definition, property, or postulate that is represented by the diagram.

   •————————•——•
   J   K   L

   $JK + KL = JL$

2. Solve $8x + 4y = 4$ for y. Justify each step.

3. Solve $4 > |x - 2|$. Graph the solution.

4. Rewrite the conditional statement in if-then form.

   A smartphone displays a warning when its battery drops below 10%.

5. You train athletes between 10 and 15 hours per week at a gym. You earn at least $10 per hour. Write a system that represents the situation.

6. Use inductive reasoning to make a conjecture about the square of an even number. Then use deductive reasoning to show that the conjecture is true.

7. Approximate when the function is positive, negative, increasing, or decreasing. Then describe the end behavior of the function.

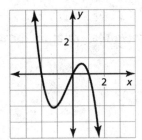

2.4 Self-Assessment

Use the scale to rate your understanding of the learning target and the success criteria.

| 1 I do not understand. | 2 I can do it with help. | 3 I can do it on my own. | 4 I can teach someone else. |

	Rating	Date
2.4 Algebraic Reasoning		
Learning Target: Use properties of equality to solve problems.	1 2 3 4	
I can identify algebraic properties of equality.	1 2 3 4	
I can use algebraic properties of equality to solve equations.	1 2 3 4	
I can use properties of equality to solve for geometric measures.	1 2 3 4	

2.5 Extra Practice

In Exercises 1 and 2, complete the proof.

1. **Given** $\overline{AB}$ and $\overline{CD}$ bisect each other at point M, and $\overline{BM} \cong \overline{CM}$.
   Prove $AB = AM + DM$

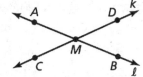

STATEMENTS	REASONS
1. $\overline{BM} \cong \overline{CM}$	1. Given
2. $\overline{CM} \cong \overline{DM}$	2. _____
3. $\overline{BM} \cong \overline{DM}$	3. _____
4. $BM = DM$	4. _____
5. _____	5. Segment Addition Postulate
6. $AB = AM + DM$	6. _____

2. **Given** $\angle AEB$ is a complement of $\angle BEC$.
   Prove $m\angle AED = 90°$

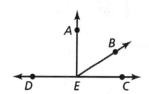

STATEMENTS	REASONS
1. $\angle AEB$ is a complement of $\angle BEC$.	1. Given
2. _____	2. Definition of complementary angles
3. $m\angle AEC = m\angle AEB + m\angle BEC$	3. _____
4. $m\angle AEC = 90°$	4. _____
5. $m\angle AED + m\angle EC = 180°$	5. Definition of supplementary angles
6. _____	6. Substitution Property of Equality
7. $m\angle AED = 90°$	7. _____

2.5 Review & Refresh

1. Solve $9x^2 - 144 = 0$ using any method. Explain your choice of method.

2. Does the table represent a *linear* or *nonlinear* function?

x	1	3	4	5
y	2	4	6	8

3. $\angle 1$ is the complement of $\angle 3$, and $m\angle 3 = 7°$. Find $m\angle 1$.

4. Use inductive reasoning to make a conjecture about the sum of an integer and the square of the integer. Then use deductive reasoning to show that the conjecture is true.

5. Solve $-2(3x + 5) = 3x + 17$. Justify each step.

6. Name the property that $\angle D \cong \angle D$ illustrates.

7. A website host charges members an initial fee of $15 and a monthly fee of $3.75. Find the total cost of the first year of membership.

8. Sketch a diagram showing $\overrightarrow{BD}$ bisecting $\angle ABC$, so that $\angle ABD \cong \angle CBD$.

2.5 Self-Assessment

Use the scale to rate your understanding of the learning target and the success criteria.

| 1 I do not understand. | 2 I can do it with help. | 3 I can do it on my own. | 4 I can teach someone else. |

	Rating	Date
2.5 Proving Statements about Segments and Angles		
Learning Target: Prove statements about segments and angles.	1 2 3 4	
I can explain the structure of a two-column proof.	1 2 3 4	
I can write a two-column proof.	1 2 3 4	
I can identify properties of congruence.	1 2 3 4	

2.6 Extra Practice

1. Complete the flowchart proof. Then write a two-column proof.

   Given ∠ACB and ∠ACD are supplementary.
   ∠EGF and ∠ACD are supplementary.

   Prove ∠ACB ≅ ∠EGF

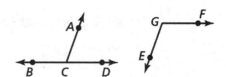

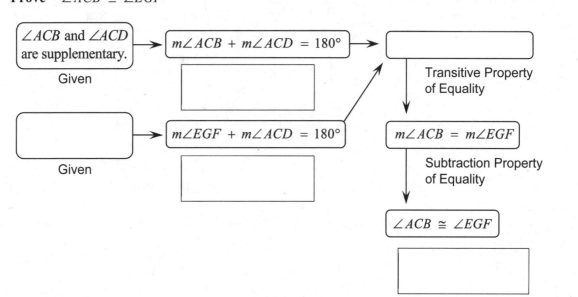

Two-Column Proof

STATEMENTS	REASONS

2. ∠ABC and ∠DBE are vertical angles. $m\angle ABC = (2x + 5)°$ and $m\angle DBE = (3x - 2)°$. Find the value of x.

Name _____ Date _____

2.6 Review & Refresh

In Exercises 1 and 2, use the rectangular prism.

1. Name three collinear points.

2. Write an example of the Three Point Postulate.

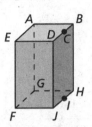

3. The final velocity v_f of an object is given by the formula $v_f = v_i + at$, where v_i is the initial velocity, a is the acceleration, and t is the time.

   a. Solve the formula for a.

   b. A vehicle with an initial velocity of 10 meters per second accelerates at a constant rate for four seconds. The final velocity of the vehicle is 15 meters per second. What is the acceleration?

4. Complete the square for $x^2 - 12x$. Then factor the trinomial.

5. Complete the two-column proof.

   Given $\angle ABD$ is a straight angle.
   $\angle CBE$ is a straight angle.

   Prove $\angle ABC \cong \angle DBE$

   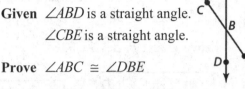

STATEMENTS	REASONS
1. $\angle ABD$ is a straight angle. $\angle CBE$ is a straight angle.	1. Given
2. $\angle ABC$ and $\angle CBD$ are supplementary.	2. _____ _____ _____ _____
3. _____ _____ _____ _____	3. Definition of supplementary angles
4. _____ _____ _____	4. Congruent Supplements Theorem

2.6 Self-Assessment

Use the scale to rate your understanding of the learning target and the success criteria.

| 1 | I do not understand. | 2 | I can do it with help. | 3 | I can do it on my own. | 4 | I can teach someone else. |

	Rating	Date
2.6 Proving Geometric Relationships		
Learning Target: Prove geometric relationships.	1 2 3 4	
I can prove geometric relationships by writing flowchart proofs.	1 2 3 4	
I can prove geometric relationships by writing paragraph proofs.	1 2 3 4	

30 Geometry
Practice Workbook and Test Prep

Name_____ Date_____

Chapter 2 — Chapter Self-Assessment

Use the scale to rate your understanding of the learning target and the success criteria.

| 1 | I do not understand. | 2 | I can do it with help. | 3 | I can do it on my own. | 4 | I can teach someone else. |

	Rating	Date
Chapter 2 Reasoning and Proofs		
Learning Target: Understand reasoning and proofs.	1 2 3 4	
I can use inductive and deductive reasoning.	1 2 3 4	
I can justify steps using algebraic reasoning.	1 2 3 4	
I can explain postulates using diagrams.	1 2 3 4	
I can prove geometric relationships.	1 2 3 4	
2.1 Conditional Statements		
Learning Target: Understand and write conditional statements.	1 2 3 4	
I can write conditional statements.	1 2 3 4	
I can write biconditional statements.	1 2 3 4	
I can determine if conditional statements are true by using truth tables.	1 2 3 4	
2.2 Inductive and Deductive Reasoning		
Learning Target: Use inductive and deductive reasoning.	1 2 3 4	
I can use inductive reasoning to make conjectures.	1 2 3 4	
I can use deductive reasoning to verify conjectures.	1 2 3 4	
I can distinguish between inductive and deductive reasoning.	1 2 3 4	
2.3 Postulates and Diagrams		
Learning Target: Interpret and sketch diagrams.	1 2 3 4	
I can identify postulates represented by diagrams.	1 2 3 4	
I can sketch a diagram given a verbal description.	1 2 3 4	
I can interpret a diagram.	1 2 3 4	

Chapter 2 Chapter Self-Assessment (continued)

	Rating	Date
2.4 Algebraic Reasoning		
Learning Target: Use properties of equality to solve problems.	1 2 3 4	
I can identify algebraic properties of equality.	1 2 3 4	
I can use algebraic properties of equality to solve equations.	1 2 3 4	
I can use properties of equality to solve for geometric measures.	1 2 3 4	
2.5 Proving Statements about Segments and Angles		
Learning Target: Prove statements about segments and angles.	1 2 3 4	
I can explain the structure of a two-column proof.	1 2 3 4	
I can write a two-column proof.	1 2 3 4	
I can identify properties of congruence.	1 2 3 4	
2.6 Proving Geometric Relationships		
Learning Target: Prove geometric relationships.	1 2 3 4	
I can prove geometric relationships by writing flowchart proofs.	1 2 3 4	
I can prove geometric relationships by writing paragraph proofs.	1 2 3 4	

Name_____ Date_____

Chapter 2 Test Prep

1. Classify a 100° angle as *acute*, *right*, *obtuse*, or *straight*.

   _____

2. Which postulate is represented by the diagram?

   Ⓐ Three Point Postulate

   Ⓑ Line-Point Postulate

   Ⓒ Plane Intersection Postulate

   Ⓓ Line Intersection Postulate

3. What is the distance from $J(-2, 1)$ to $K(2, 1)$ in the coordinate plane?

   Ⓐ −4

   Ⓑ 0

   Ⓒ 2

   Ⓓ 4

4. Find the sum of the next three terms of the arithmetic sequence 8, 13, 18, 23, ….

5. ∠3 and ∠4 are vertical angles, and $m\angle 4 = 11.5°$. What is $m\angle 3$? _____ degrees

Chapter 2 Test Prep (continued)

6. Which statement has the same truth value as $p \rightarrow q$?

   Ⓐ $\sim p \rightarrow \sim q$

   Ⓑ $p \rightarrow \sim q$

   Ⓒ $\sim q \rightarrow \sim p$

   Ⓓ $q \rightarrow \sim p$

7. What is the value of x?

   Ⓐ -4

   Ⓑ -2

   Ⓒ 2

   Ⓓ 4

   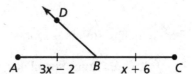

8. Select all the polynomials that are factored *completely*.

   Ⓐ $4x^2 + 2x = 2(2x^2 + x)$

   Ⓑ $x^4 - y^4 = (x^2 + y^2)(x^2 - y^2)$

   Ⓒ $3x + 6 = 3(x + 2)$

   Ⓓ $z^3 + 3z^2 + 6z + 9 = z^2(z + 3) + 6(z + 3)$

   Ⓔ $2w^4 + 4w^3 + w + 2 = (2w^3 + 1)(w + 2)$

9. What property is demonstrated below?

   If $m\angle 1 = m\angle 2$, then $m\angle 1 + m\angle 3 = m\angle 2 + m\angle 3$.

   _____

10. What is the inverse of $f(x) = \sqrt{x - 4}$?

    Ⓐ $f^{-1}(x) = 2x + 4$

    Ⓑ $f^{-1}(x) = x^2 + 4, x \geq 0$

    Ⓒ $f^{-1}(x) = x^2 - 4, x \geq 2$

    Ⓓ $f^{-1}(x) = x^2 + 4$

Chapter 2 Test Prep (continued)

11. Which is a counterexample to the conjecture that the sum of three consecutive whole numbers is even?

Ⓐ $-4 + (-3) + (-2) = -9$

Ⓑ $1 + 2 + 3 = 6$

Ⓒ $3 + 5 + 7 = 15$

Ⓓ $10 + 11 + 12 = 33$

12. Which reason corresponds with the fourth statement in the proof, "$m\angle ABD + m\angle DBE = m\angle CBE + m\angle DBE$?"

Ⓐ Substitution Property of Equality

Ⓑ Distributive Property

Ⓒ Transitive Property of Equality

Ⓓ Reflexive Property of Equality

Given $m\angle ABE = m\angle CBD$
Prove $m\angle ABD = m\angle CBE$

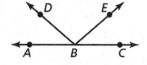

STATEMENTS	REASONS
1. $m\angle ABE = m\angle CBD$	1. Given
2. $m\angle ABE = m\angle ABD + m\angle DBE$	2. Angle Addition Postulate
3. $m\angle CBD = m\angle CBE + m\angle DBE$	3. Angle Addition Postulate
4. $m\angle ABD + m\angle DBE = m\angle CBE + m\angle DBE$	4.
5. $m\angle ABD + m\angle DBE - m\angle DBE = m\angle CBE + m\angle DBE - m\angle DBE$	5. Subtraction Property of Equality
6. $m\angle ABD = m\angle CBE$	6. Simplify.

13. Select all true equations.

Ⓐ $5^0 \stackrel{?}{=} 0$

Ⓑ $5^{-3} \cdot 5^8 \stackrel{?}{=} 5^5$

Ⓒ $\dfrac{5^3}{5^5} \stackrel{?}{=} 5^2$

Ⓓ $(5^2)^3 \stackrel{?}{=} 5^6$

Ⓔ $\dfrac{5^{-4}}{5^3} \stackrel{?}{=} \dfrac{1}{5^7}$

14. Which point is the remaining vertex of a triangle with $A(-2, 1)$ and $B(-2, -5)$ that has an area of 6 square units?

Ⓐ $C(0, 1)$

Ⓑ $C(3, -2)$

Ⓒ $C(-1, 1)$

Ⓓ $C(4, 1)$

Chapter 2 Test Prep (continued)

15. Describe and graph the inequality $-1 \leq x \leq 2$.

Ⓐ four points;

Ⓑ line segment;

Ⓒ two rays;

Ⓓ line that passes through -1 and 2;

16. Rewrite the perimeter formula $P = r(\theta + 2)$ in terms of θ.

Ⓐ $\theta = P - r - 2$

Ⓑ $\theta = \dfrac{P}{r - 2}$

Ⓒ $\theta = \dfrac{P}{r} - 2$

Ⓓ $\theta = 2P - r$

17. Rewrite the conditional statement in if-then form. You cannot live chat when you log off.

Ⓐ If you log off, then you cannot live chat.

Ⓑ If you log off, then you can live chat.

Ⓒ If you cannot live chat, then you cannot log off.

Ⓓ If you can live chat, then you log off.

18. A drawbridge is 22 degrees downward from a vertical position. The angle increases at a rate of 2 degrees per second. How long will it take the bridge to be horizontal?

_____ seconds

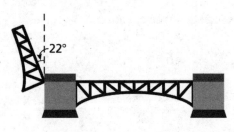

Name_____ Date_____

3.1 Extra Practice

In Exercises 1–4, consider the lines that contain the segments in the figure and the planes that contain the faces of the figure. Which line(s) or plane(s) contain point B and appear to fit the description?

1. line(s) skew to $\overrightarrow{FG}$

2. line(s) perpendicular to $\overrightarrow{FG}$

3. line(s) parallel to $\overrightarrow{FG}$

4. plane(s) parallel to plane FGH

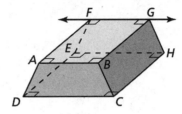

In Exercises 5–8, use the diagram.

5. Name a pair of parallel lines.

6. Name a pair of perpendicular lines.

7. Is $\overrightarrow{WX} \parallel \overrightarrow{QR}$? Explain.

8. Is $\overrightarrow{ST} \perp \overrightarrow{NV}$? Explain.

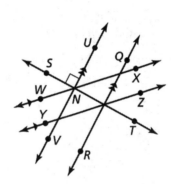

In Exercises 9–12, identify all pairs of angles of the given type.

9. corresponding

10. alternate interior

11. alternate exterior

12. consecutive interior

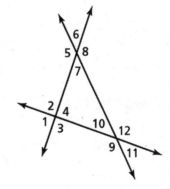

13. How many pairs of consecutive interior angles do you have when two coplanar lines are intersected by a transversal? How many pairs of consecutive interior angles do you have when three coplanar lines are intersected by a transversal? How many pairs of consecutive interior angles do you have when n coplanar lines are intersected by a transversal?

3.1 Review & Refresh

1. Copy the segment and construct a segment bisector by paper folding. Then label the midpoint M.

2. Solve the inequality $x + 8 < 13$. Graph the solution.

3. Name the property that the statement illustrates. If $\angle K \cong \angle L$, then $\angle L \cong \angle K$.

4. Classify the pair of numbered angles.

   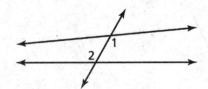

5. Solve the system.

   $y = \frac{1}{2}x + 3$

   $y = \frac{1}{4}x + 1$

6. Use the Transitive Property of Segment Congruence to complete the statement. If $\overline{WX} \cong \overline{YZ}$ and $\overline{YZ} \cong \overline{QR}$, then _____.

7. Write a proof using any format.
   Given M is the midpoint of $\overline{AB}$.
   $\overline{CM} \cong \overline{MB}$
   Prove $\overline{AM} \cong \overline{CM}$

   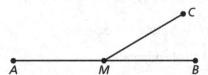

8. Write an equation of the line that passes through the point $(4, -2)$ and has a slope of -3.

9. Evaluate $\sqrt[3]{-27}$.

10. Find the volume of a cylinder with a radius of 7 meters and a height of 6 meters. Round your answer to the nearest tenth.

3.1 Self-Assessment

Use the scale to rate your understanding of the learning target and the success criteria.

| 1 I do not understand. | 2 I can do it with help. | 3 I can do it on my own. | 4 I can teach someone else. |

	Rating	Date
3.1 Pairs of Lines and Angles		
Learning Target: Understand lines, planes, and pairs of angles.	1 2 3 4	
I can identify lines and planes.	1 2 3 4	
I can identify parallel and perpendicular lines.	1 2 3 4	
I can identify pairs of angles formed by transversals.	1 2 3 4	

Name_____ Date_____

3.2 Extra Practice

In Exercises 1–4, find $m\angle 1$ and $m\angle 2$. Tell which theorem you use in each case.

1.
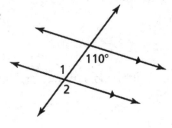

2.
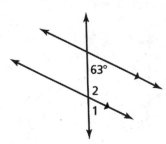

3.
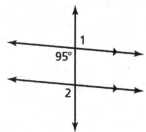

4.
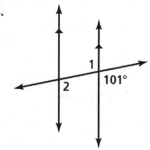

In Exercises 5 and 6, find the value of x. Show your steps.

5.
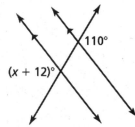

6.
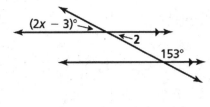

7. Prove that if $\angle 1 \cong \angle 2$, then $\angle 2 \cong \angle 3$. What is $m\angle 1$? Explain.

3.2 Review & Refresh

In Exercises 1–3, use the diagram.

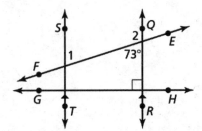

1. Name a pair of perpendicular lines.

2. Is $\overrightarrow{EF} \parallel \overrightarrow{GH}$? Explain.

3. Find $m\angle 1$ and $m\angle 2$. Tell which postulates or theorems you used.

In Exercises 4 and 5, name the property that the statement illustrates.

4. If $\overline{LM} \cong \overline{PR}$, then $\overline{PR} \cong \overline{LM}$.

5. If $\angle W \cong \angle X$ and $\angle X \cong \angle Y$, then $\angle W \cong \angle Y$.

6. Factor the polynomial completely.
$16x^6 - 64x^4$

7. Find the x- and y-intercepts of the graph of $3y - 8x = 24$.

8. A square painting is surrounded by a frame with uniform width. The painting has a side length of $(x - 2)$ inches. The side length of the frame is $(x + 1)$ inches. Write an expression for the area of the square frame. Then find the area of the frame when $x = 6$.

9. Find the value of x in the diagram.

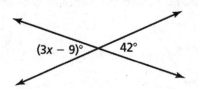

3.2 Self-Assessment

Use the scale to rate your understanding of the learning target and the success criteria.

| 1 I do not understand. | 2 I can do it with help. | 3 I can do it on my own. | 4 I can teach someone else. |

	Rating	Date
3.2 Parallel Lines and Transversals		
Learning Target: Prove and use theorems about parallel lines.	1 2 3 4	
I can use properties of parallel lines to find angle measures.	1 2 3 4	
I can prove theorems about parallel lines.	1 2 3 4	

3.3 Extra Practice

In Exercises 1 and 2, find the value of x that makes m ∥ n. Explain your reasoning.

1.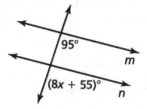

2.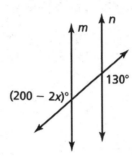

In Exercises 3–6, decide whether there is enough information to prove that m ∥ n. If so, state the theorem you can use.

3.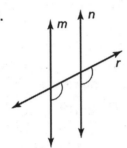

4.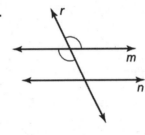

5.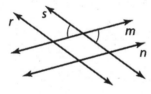

6.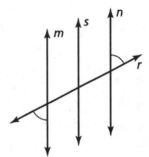

7. The angles formed between the braces and the wings of a biplane are shown in the figure. Are the top and bottom wings of a biplane parallel? Explain your reasoning.

Name_____ Date _____

3.3 Review & Refresh

In Exercises 1 and 2, find the distance between the two points.

1. $(8, 6)$ and $(0, -9)$ 2. $(11, -2)$ and $(5, 3)$

3. Find the value of x.

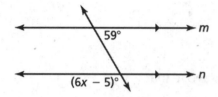

4. The height (in feet) of a ball t seconds after it is thrown can be represented by $h(t) = -16t^2 + 128t + 7$. Estimate and interpret the maximum value of the function.

5. Find the value of x that makes $k \parallel \ell$. Explain your reasoning.

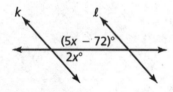

In Exercises 6 and 7, use the diagram.

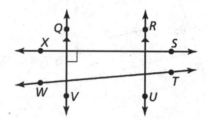

6. Name a pair of perpendicular lines.

7. Is $\overrightarrow{XS} \parallel \overrightarrow{WT}$? Explain.

In Exercises 8 and 9, solve the system using any method. Explain your choice of method.

8. $3x + 4y = 12$
   $-6x - 5y = 30$

9. $y = 9x + 7$
   $4x - y = -17$

10. Evaluate $f(x) = -8x + 14$ when $x = -3, 2,$ and 6.

3.3 Self-Assessment

Use the scale to rate your understanding of the learning target and the success criteria.

| 1 I do not understand. | 2 I can do it with help. | 3 I can do it on my own. | 4 I can teach someone else. |

	Rating	Date
3.3 Proofs with Parallel Lines		
Learning Target: Prove and use theorems about identifying parallel lines.	1 2 3 4	
I can use theorems to identify parallel lines.	1 2 3 4	
I can construct parallel lines.	1 2 3 4	
I can prove theorems about identifying parallel lines.	1 2 3 4	

Name_____ Date_____

3.4 Extra Practice

In Exercises 1 and 2, find the distance from point A to $\overrightarrow{BC}$.

1.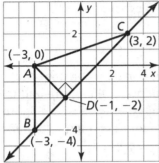

2.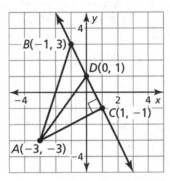

In Exercises 3–6, determine which lines, if any, must be parallel. Explain your reasoning.

3.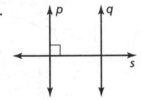

4.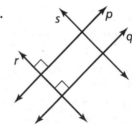

5.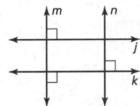

6.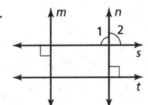

7. Your friend claims that there is only one line that can be drawn perpendicular to $\overline{PQ}$. Is your friend correct? Explain your reasoning.

8. Determine which lines must be parallel. Explain your reasoning.

Name_____ Date_____

3.4 Review & Refresh

In Exercises 1 and 2, find the slope and the y-intercept of the graph of the linear equation.

1. $y = -\frac{2}{3}x + 1$
2. $7x + y = 16$

3. Two angles form a linear pair. The measure of one angle is 94°. Find the measure of the other angle.

4. Find the slope of the line that passes through $(-3, 8)$ and $(1, 6)$.

5. The post office and the bank are both on the same straight road between the school and your house. The distance from the school to the bank is 523 yards, the distance from the bank to your house is 803 yards, and the distance from the post office to your house is 391 yards.

   a. What is the distance from the post office to the bank?

   b. What is the distance from the school to your house?

6. Solve the system using any method.
$$y = x^2 + 3x - 1$$
$$y = -2x - 5$$

In Exercises 7–9, consider the lines that contain the segments in the figure and the planes that contain the faces of the figure. Which line(s) or plane(s) contain point *M* and appear to fit the description?

7. line(s) perpendicular to $\overrightarrow{JK}$

8. line(s) skew to $\overrightarrow{JK}$

9. plane(s) parallel to plane *QRK*

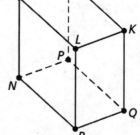

In Exercises 10 and 11, graph the function. Compare the graph to the graph of $f(x) = |x|$. Find the domain and range.

10. $g(x) = |x| - 7$
11. $h(x) = -\frac{3}{4}|x|$

3.4 Self-Assessment

Use the scale to rate your understanding of the learning target and the success criteria.

| 1 I do not understand. | 2 I can do it with help. | 3 I can do it on my own. | 4 I can teach someone else. |

	Rating	Date
3.4 Proofs with Perpendicular Lines		
Learning Target: Prove and use theorems about perpendicular lines.	1 2 3 4	
I can find the distance from a point to a line.	1 2 3 4	
I can construct perpendicular lines and perpendicular bisectors.	1 2 3 4	
I can prove theorems about perpendicular lines.	1 2 3 4	

Name_____ Date_____

3.5 Extra Practice

In Exercises 1 and 2, find the coordinates of point P along the directed line segment AB so that AP to PB is the given ratio.

1. $A(-2, 7)$, $B(-4, 1)$; 3 to 1

2. $A(3, 1)$, $B(8, -2)$; 2 to 3

3. Determine which of the lines are parallel and which of the lines are perpendicular.

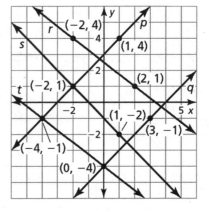

4. Tell whether the lines through the given points are *parallel*, *perpendicular*, or *neither*. Justify your answer.

   Line 1: $(2, 0), (-2, 2)$
   Line 2: $(1, -2), (4, 4)$

5. Write an equation of the line passing through the point $(3, -2)$ that is parallel to the line $y = \frac{2}{3}x - 1$. Graph the equations of the lines to check that they are parallel.

6. Write an equation of the line passing through the point $(-2, 2)$ that is perpendicular to the line $y = 2x + 3$. Graph the equations of the lines to check that they are perpendicular.

7. Find the distance from the point $(0, 5)$ to the line $y = -3x - 5$.

Name _____ Date _____

3.5 Review & Refresh

1. Find the value of x that makes $m \parallel n$. Explain your reasoning.

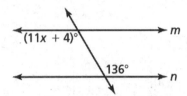

2. Make and test a conjecture about the product of three consecutive even numbers.

3. Find the perimeter of the triangle with the vertices $(2, 0)$, $(8, 5)$, and $(2, 7)$.

4. Solve the equation $\left(\frac{1}{3}\right)^x = 81$. Check your solution.

5. Determine which lines, if any, must be parallel. Explain your reasoning.

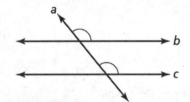

6. Factor $6x^2 + 7x + 2$.

7. Write an equation of the line passing through point $P(4, -3)$ that is parallel to $y = -2x + 9$.

8. Find the domain of $p(x) = \sqrt{8x + 3}$.

9. Solve $x^2 + 18x = 0$.

10. A chute forms a line between two parallel supports, as shown. Find $m\angle 2$. Explain your reasoning.

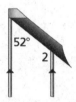

11. Solve the inequality $4w + 11 \leq 5w + 2w - 1$. Graph the solution.

12. Graph the quadratic function $y = (x - 3)(x + 1)$. Label the vertex, axis of symmetry, and x-intercepts. Find the domain and range of the function.

3.5 Self-Assessment

Use the scale to rate your understanding of the learning target and the success criteria.

| 1 I do not understand. | 2 I can do it with help. | 3 I can do it on my own. | 4 I can teach someone else. |

	Rating	Date
3.5 Equations of Parallel and Perpendicular Lines		
Learning Target: Partition a directed line segment and understand slopes of parallel and perpendicular lines.	1 2 3 4	
I can partition directed line segments using slope.	1 2 3 4	
I can use slopes to identify parallel and perpendicular lines.	1 2 3 4	
I can write equations of parallel and perpendicular lines.	1 2 3 4	
I can find the distance from a point to a line.	1 2 3 4	

Name_____ Date_____

 Chapter Self-Assessment

Use the scale to rate your understanding of the learning target and the success criteria.

1 I do not understand. **2** I can do it with help. **3** I can do it on my own. **4** I can teach someone else.

	Rating	Date
Chapter 3 Parallel and Perpendicular Lines		
Learning Target: Understand parallel and perpendicular lines.	1 2 3 4	
I can identify lines and angles.	1 2 3 4	
I can describe angle relationships formed by parallel lines and a transversal.	1 2 3 4	
I can prove theorems involving parallel and perpendicular lines.	1 2 3 4	
I can write equations of parallel and perpendicular lines.	1 2 3 4	
3.1 Pairs of Lines and Angles		
Learning Target: Understand lines, planes, and pairs of angles.	1 2 3 4	
I can identify lines and planes.	1 2 3 4	
I can identify parallel and perpendicular lines.	1 2 3 4	
I can identify pairs of angles formed by transversals.	1 2 3 4	
3.2 Parallel Lines and Transversals		
Learning Target: Prove and use theorems about parallel lines.	1 2 3 4	
I can use properties of parallel lines to find angle measures.	1 2 3 4	
I can prove theorems about parallel lines.	1 2 3 4	
3.3 Proofs with Parallel Lines		
Learning Target: Prove and use theorems about identifying parallel lines.	1 2 3 4	
I can use theorems to identify parallel lines.	1 2 3 4	
I can construct parallel lines.	1 2 3 4	
I can prove theorems about identifying parallel lines.	1 2 3 4	

Name _____ Date _____

Chapter Self-Assessment (continued)

	Rating	Date
3.4 Proofs with Perpendicular Lines		
Learning Target: Prove and use theorems about perpendicular lines.	1 2 3 4	
I can find the distance from a point to a line.	1 2 3 4	
I can construct perpendicular lines and perpendicular bisectors.	1 2 3 4	
I can prove theorems about perpendicular lines.	1 2 3 4	
3.5 Equations of Parallel and Perpendicular Lines		
Learning Target: Partition a directed line segment and understand slopes of parallel and perpendicular lines.	1 2 3 4	
I can partition directed line segments using slope.	1 2 3 4	
I can use slopes to identify parallel and perpendicular lines.	1 2 3 4	
I can write equations of parallel and perpendicular lines.	1 2 3 4	
I can find the distance from a point to a line.	1 2 3 4	

Chapter 3 Test Prep

1. Which of the following statements *cannot* be assumed from the diagram?

    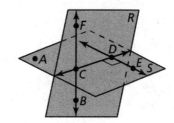

    Ⓐ Points B, C, and F are collinear.

    Ⓑ $\overline{CF} \perp$ plane S

    Ⓒ $\overline{DE} \perp$ plane R

    Ⓓ $\overrightarrow{FB}$ intersects $\overline{CD}$ at point C.

2. Identify the contrapositive of the given statement.
    Given Statement
    If two angles form a linear pair, then the angles are supplementary.

    Ⓐ If two angles do not form a linear pair, then the angles are not supplementary.

    Ⓑ If two angles are supplementary, then the angles form a linear pair.

    Ⓒ Two angles form a linear pair if and only if the angles are supplementary.

    Ⓓ If two angles are not supplementary, then the angles do not form a linear pair.

3. What is the value of x?

    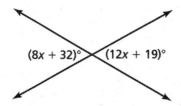

    x =

4. What is the value of x?

    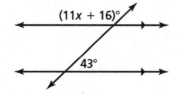

    x =

Name _____ Date _____

Chapter 3 Test Prep (continued)

5. Classify the pair of numbered angles.

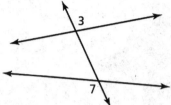

Ⓐ alternate interior angles
Ⓑ alternate exterior angles
Ⓒ corresponding angles
Ⓓ consecutive interior angles

6. What is the value of x that makes $m \parallel n$?

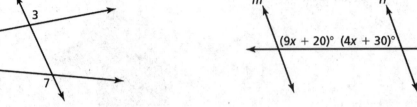

Ⓐ 2
Ⓑ 3
Ⓒ 10
Ⓓ 38

7. Point M is the midpoint of $\overline{AB}$. Find AB.

A •―――3y − 1―――• M •―――2y + 10―――• B

Ⓐ 64
Ⓑ 11
Ⓒ 16
Ⓓ 32

8. What is the value of y that makes $t \parallel u$?

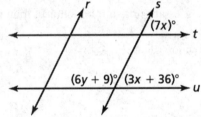

Ⓐ 18
Ⓑ 9
Ⓒ 117
Ⓓ 63

9. Select all the pairs of lines that must be parallel.

Ⓐ $a \parallel c$
Ⓑ $b \parallel c$
Ⓒ $d \parallel e$
Ⓓ $a \parallel b$

10. Write an equation of the line passing through the point $(12, -7)$ that is perpendicular to the line $y = 3x + 1$.

Chapter 3 Test Prep (continued)

11. Use the Law of Syllogism to write a new conditional statement that follows from the pair of true statements.

If a figure is a square, then the figure has exactly four congruent sides. If a figure has exactly four congruent sides, then the figure is a rhombus.

Ⓐ If a figure is a rhombus, then the figure is a square.

Ⓑ A figure is a square if and only if the figure is a rhombus.

Ⓒ If a figure is a square, then the figure is a rhombus.

Ⓓ If a figure is not a square, then the figure is not a rhombus.

12. $\angle XYZ$ and $\angle LMN$ are complementary angles. $m\angle XYZ = (3x + 14)°$ and $m\angle LMN = (5x + 24)°$. What is $m\angle LMN$?

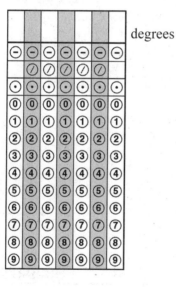 degrees

13. Find the coordinates of point P along the directed line segment AB so that the ratio of AP to PB is 1 to 3.

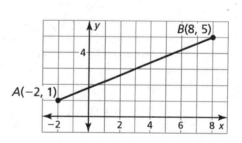

14. Which of the following is the remaining vertex of a triangle that has an area of 8 square units?

Ⓐ $R(0, 3)$

Ⓑ $S(-4, 5)$

Ⓒ $T(2, 3)$

Ⓓ $U(-4, 2)$

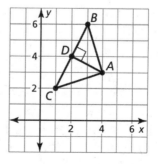

Chapter 3 Test Prep (continued)

15. Consider the lines that contain the segments in the figure and the planes that contain the faces of the figure. Select all the lines that appear parallel to $\overleftrightarrow{BC}$.

Ⓐ $\overleftrightarrow{AB}$

Ⓑ $\overleftrightarrow{AD}$

Ⓒ $\overleftrightarrow{EF}$

Ⓓ $\overleftrightarrow{HE}$

Ⓔ $\overleftrightarrow{DG}$

Ⓕ $\overleftrightarrow{HG}$

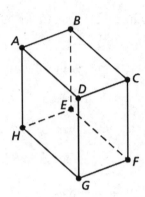

16. What is the distance from point A to $\overleftrightarrow{BC}$?

Ⓐ $5\sqrt{2}$

Ⓑ $\sqrt{5}$

Ⓒ $4\sqrt{2}$

Ⓓ 5

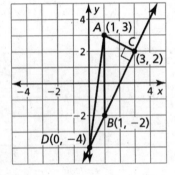

17. Which of the following statements is false?

Ⓐ Two planes intersect in a line.

Ⓑ Two points determine a line.

Ⓒ Two lines intersect in a point.

Ⓓ Any three points determine a plane.

18. Which reason corresponds with the fifth statement in the proof, "$EF = GH$?"

Ⓐ Subtraction Property of Equality

Ⓑ Segment Addition Postulate

Ⓒ Substitution Property of Equality

Ⓓ Reflexive Property of Equality

Given $EG = FH$

Prove $EF = GH$

STATEMENTS	REASONS
1. $EG = FH$	1. Given
2. $EF + FG = EG$	2. Segment Addition Postulate
3. $FG + GH = FH$	3. Segment Addition Postulate
4. $EF + FG = FG + GH$	4. Substitution Property of Equality
5. $EF = GH$	5. _____

Name_____ Date_____

4.1 Extra Practice

In Exercises 1–3, name the vector and write its component form.

1.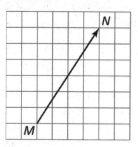
2.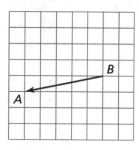
3.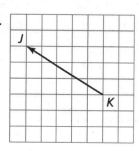

In Exercises 4 and 5, the vertices of △ABC are A(1, 2), B(5, 1) and C(5, 4). Translate △ABC using the given vector. Graph △ABC and its image.

4. $\langle -4, 0 \rangle$

5. $\langle -2, -4 \rangle$

In Exercises 6 and 7, write a rule for the translation of quadrilateral PQRS to quadrilateral P'Q'R'S'.

6.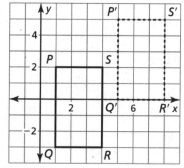

7.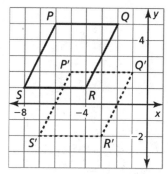

8. Graph $\overline{AB}$ with endpoints $A(3, -1)$ and $B(-2, 0)$ and its image after the composition.

   Translation: $(x, y) \rightarrow (x - 2, y + 3)$
   Translation: $(x, y) \rightarrow (x + 1, y - 1)$

9. In a video game, you move a spaceship 1 unit left and 4 units up. Then, you move the spaceship 2 units left. Rewrite the composition as a single translation.

Name _____ Date _____

4.1 Review & Refresh

1. Decide whether there is enough information to prove that $m \parallel n$. If so, state the theorem you can use.

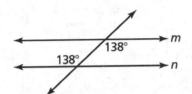

2. Write an equation of the line passing through point $P(-6, 4)$ that is perpendicular to $y + 8 = 2(x - 14)$. Graph the equations to check that the lines are perpendicular.

3. Graph quadrilateral $ABCD$ with vertices $A(-2, 4)$, $B(-1, 6)$, $C(4, 4)$, and $D(2, 3)$ and its image after the translation $(x, y) \rightarrow (x + 2, y - 3)$.

4. Write an equation for the nth term of the arithmetic sequence. Then find a_{10}.

   7, 3, −1, −5, …

5. Solve the equation $50x = 2x^3$.

6. Graph the function $g(x) = -1.5x^2$. Compare the graph to the graph of $f(x) = x^2$.

7. The function $h(x) = -16x^2 + 32x + 4$ models the height h (in feet) of a ball ejected from a ball launcher after x seconds. When is the ball at a height of 15 feet?

8. Find the inverse of the function $f(x) = -\frac{3}{5}x + \frac{9}{5}$. Then graph the function and its inverse.

4.1 Self-Assessment

Use the scale to rate your understanding of the learning target and the success criteria.

| 1 I do not understand. | 2 I can do it with help. | 3 I can do it on my own. | 4 I can teach someone else. |

	Rating	Date
4.1 Translations		
Learning Target: Understand translations of figures.	1 2 3 4	
I can translate figures.	1 2 3 4	
I can write a translation rule for a given translation.	1 2 3 4	
I can explain what a rigid motion is.	1 2 3 4	
I can perform a composition of translations on a figure.	1 2 3 4	

4.2 Extra Practice

In Exercises 1 and 2, graph the polygon with the given vertices and its image after a reflection in the given line.

1. $A(-1, 5)$, $B(-4, 4)$, $C(-3, 1)$; y-axis

2. $A(0, 2)$, $B(4, 5)$, $C(5, 2)$; x-axis

In Exercises 3 and 4, graph the image of the polygon after a reflection in the given line.

3. $y = x$

4. $y = -x$

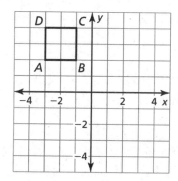
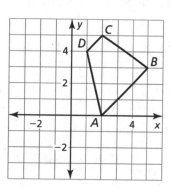

In Exercises 5 and 6, graph $\triangle JKL$ **with vertices** $J(3, 1)$, $K(4, 2)$, **and** $L(1, 3)$ **and its image after the glide reflection.**

5. Translation: $(x, y) \to (x - 6, y - 1)$
   Reflection: in the line $y = -x$

6. Translation: $(x, y) \to (x, y - 4)$
   Reflection: in the line $x = 1$

In Exercises 7–10, identify the line symmetry (if any) of the word.

7. MOON

8. WOW

9. KID

10. DOCK

11. The line $y = 3x - 5$ is reflected in the line $x = a$ so that its image is given by $y = 1 - 3x$. What is the value of a?

4.2 Review & Refresh

1. Find the distance from point A to $\overline{XZ}$.

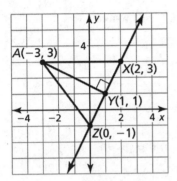

In Exercises 2 and 3, solve the equation by graphing.

2. $|3x| = |x - 4|$

3. $-2x + 8 = 3(x + 1)$

4. Graph $\triangle ABC$ with vertices $A(2, 3)$, $B(3, -1)$, and $C(-1, 2)$ and its image after a reflection in the x-axis.

5. Name the property that "If $\angle X \cong \angle Y$, then $\angle Y \cong \angle X$" illustrates.

6. Make a scatter plot of the data. Then describe the relationship between the data.

x	0.2	0.5	0.8	1.0	1.4	1.7	1.9	2.4
y	3.1	2.9	2.8	2.6	2.1	1.9	1.6	1.4

7. Find the distance from the point $(3, 7)$ to the line $y = \frac{1}{2}x + 3$.

8. Evaluate $h(x) = 4x - 9$ when $x = -2$.

9. Use the translation $(x, y) \to (x + 3, y - 2)$ to find the image of $A(-1, 6)$.

10. Factor $-4t^2 + 16t - 15$.

11. Find the product of $x - 8$ and $x + 10$.

4.2 Self-Assessment

Use the scale to rate your understanding of the learning target and the success criteria.

| 1 | I do not understand. | 2 | I can do it with help. | 3 | I can do it on my own. | 4 | I can teach someone else. |

	Rating	Date
4.2 Reflections		
Learning Target: Understand reflections of figures.	1 2 3 4	
I can reflect figures.	1 2 3 4	
I can perform compositions with reflections.	1 2 3 4	
I can identify line symmetry in polygons.	1 2 3 4	

Name_____ Date_____

4.3 Extra Practice

In Exercises 1–3, graph the image of the polygon after a rotation of the given number of degrees about the origin.

1. 180°

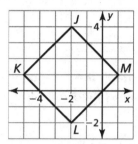

2. 90°

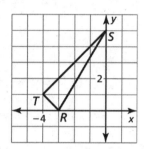

3. 270°

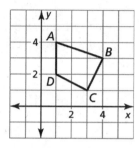

In Exercises 4 and 5, graph the image of $\overline{MN}$ after the composition.

4. **Reflection:** in the x-axis
   Rotation: 180° about the origin

5. **Rotation:** 90° about the origin
   Translation: $(x, y) \rightarrow (x + 2, y - 3)$

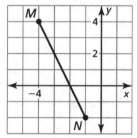
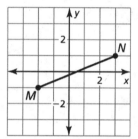

In Exercises 6 and 7, graph $\triangle JKL$ with vertices $J(2, 3)$, $K(1, -1)$, and $L(-1, 0)$ and its image after the composition.

6. **Rotation:** 180° about the origin
   Reflection: in the line $x = 2$

7. **Translation:** $(x, y) \rightarrow (x - 4, y - 4)$
   Rotation: 270° about the origin

In Exercises 8 and 9, determine whether the figure has rotational symmetry. If so, describe any rotations that map the figure onto itself.

8.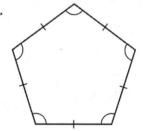

9.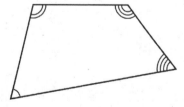

10. List the uppercase letters of the alphabet that have rotational symmetry. Describe any rotations that map the letter onto itself.

Copyright © Big Ideas Learning, LLC
All rights reserved.

Geometry
Practice Workbook and Test Prep

Name _____ Date _____

4.3 Review & Refresh

1. $\overrightarrow{DF}$ bisects $\angle CDE$. If $m\angle CDF = 61°$, find $m\angle EDF$ and $m\angle CDE$.

2. The figures are congruent. Name the corresponding angles and the corresponding sides.

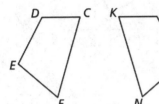

3. The endpoints of the directed line segment AB are $A(7, 4)$ and $B(2, 9)$. Find the coordinates of point P along segment AB so that the ratio of AP to PB is 3 to 2.

4. Graph the system. Identify a solution.

   $y > -x + 2$

   $y \leq \frac{4}{5}x - 1$

5. Determine whether the table represents a *linear* or an *exponential* function. Explain.

x	−2	−1	0	1	2
f(x)	1	5	9	13	17

In Exercises 6 and 7, graph the polygon with the given vertices and its image after the indicated transformation.

6. $A(3, -2)$, $B(1, 5)$, $C(-2, 0)$

   Reflection: in the y-axis

7. $W(0, 3)$, $X(-2, 1)$, $Y(-3, 3)$, $Z(-1, 5)$

   Rotation: 270° about the origin

8. Railways use railroad ties to secure the rails in place, as shown in the diagram. Each tie is parallel to the tie directly next to it. Explain why the leftmost tie is parallel to the rightmost tie.

4.3 Self-Assessment

Use the scale to rate your understanding of the learning target and the success criteria.

| 1 I do not understand. | 2 I can do it with help. | 3 I can do it on my own. | 4 I can teach someone else. |

	Rating	Date
4.3 Rotations		
Learning Target: Understand rotations of figures.	1 2 3 4	
I can rotate figures.	1 2 3 4	
I can perform compositions with rotations.	1 2 3 4	
I can identify rotational symmetry in polygons.	1 2 3 4	

Name_____ Date_____

4.4 Extra Practice

1. Identify any congruent figures in the coordinate plane. Explain.

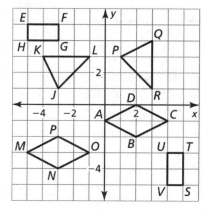

2. Describe a congruence transformation that maps polygon *ABCD* to polygon *EFGH*.

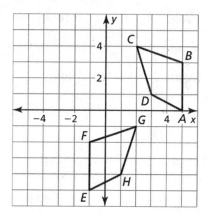

In Exercises 3 and 4, determine whether the polygons with the given vertices are congruent. Use transformations to explain your reasoning.

3. $A(2, 2)$, $B(3, 1)$, $C(1, 1)$ and
   $D(2, -2)$, $E(3, -1)$, $F(1, -1)$

4. $G(3, 3)$, $H(2, 1)$, $I(6, 2)$, $J(6, 3)$ and
   $K(-2, -1)$, $L(-3, -3)$, $M(2, -2)$, $N(2, -1)$

In Exercises 5–7, $k \parallel m$, $\overline{UV}$ is reflected in line k, and $\overline{U'V'}$ is reflected in line m.

5. Which lines are perpendicular to $\overline{UU''}$?

6. Why is V'' the image of V?

7. If the distance between k and m is 5 inches, what is the length of $\overline{VV''}$?

8. Describe a single transformation that maps A to A''.

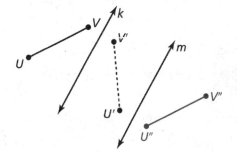

4.4 Review & Refresh

In Exercises 1 and 2, solve the equation.

1. $-3(x - 4) = -7x$
2. $3n + 1 = \frac{1}{2}(8n - 4)$

3. Yesterday, you earned $40 in tips. Today, you earned $55 in tips. What is the percent of change?

4. Describe a congruence transformation that maps polygon $ABCD$ to polygon $QRST$.

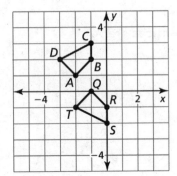

In Exercises 5 and 6, graph the linear equation. Identify the x-intercept.

5. $y = -2x + 7$
6. $4x - 2y = 16$

7. Write an inequality that represents the graph.

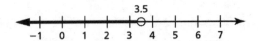

8. Let p be "you bake cookies" and let q be "you turn the oven on." Write the conditional statement $p \to q$, the converse $q \to p$, the inverse $\sim p \to \sim q$, and the contrapositive $\sim q \to \sim p$ in words. Then decide whether each statement is *true* or *false*.

In Exercises 9 and 10, find the value of x. Show your steps.

9.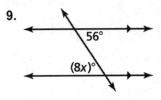

10.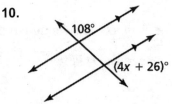

4.4 Self-Assessment

Use the scale to rate your understanding of the learning target and the success criteria.

| 1 I do not understand. | 2 I can do it with help. | 3 I can do it on my own. | 4 I can teach someone else. |

	Rating	Date
4.4 Congruence and Transformations		
Learning Target: Understand congruence transformations.	1 2 3 4	
I can identify congruent figures.	1 2 3 4	
I can describe congruence transformations.	1 2 3 4	
I can use congruence transformations to solve problems.	1 2 3 4	

4.5 Extra Practice

In Exercises 1–3, find the scale factor of the dilation. Then tell whether the dilation is a *reduction* or an *enlargement*.

1.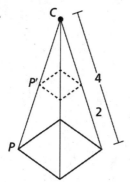

2.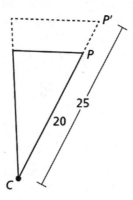

3.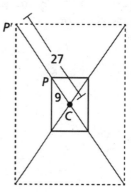

In Exercises 4 and 5, graph the polygon with the given vertices and its image after a dilation with scale factor *k*.

4. $A(-3, 1)$, $B(-4, -1)$, $C(-2, -1)$; $k = 2$

5. $P(-10, 0)$, $Q(-5, 0)$, $R(0, 5)$, $S(-5, 5)$; $k = \frac{1}{5}$

6. You design a poster on an 8.5-inch by 11-inch paper for a contest at your school. The poster of the winner will be printed on a 34-inch by 44-inch canvas to be displayed. What is the scale factor of this dilation?

7. A biology book shows the image of an insect that is 10 times its actual size. The image of the insect is 8 centimeters long. What is the actual length of the insect?

8. The old film-style cameras created photos that were best printed at 3.5 inches by 5 inches. Today's new digital cameras create photos that are best printed at 4 inches by 6 inches. Neither size photo will enlarge perfectly to fit in an 11-inch by 14-inch frame. With which type of camera will you minimize the loss of the edges of an enlarged photo?

Name _____ Date _____

4.5 Review & Refresh

In Exercises 1 and 2, graph the polygon with the given vertices and its image after the indicated transformation.

1. $D(-2, 1)$, $E(-3, 6)$, $F(1, 5)$, $G(2, 0)$
   Reflection: in the x-axis

2. $A(1, 3)$, $B(-1, 2)$, $C(2, 1)$
   Rotation: $180°$ about the origin

5. Describe a congruence transformation that maps $\triangle ABC$ to $\triangle XYZ$.

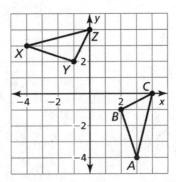

6. Graph $g(x) = 2(x + 3)^2 - 1$. Compare the graph to the graph of $f(x) = x^2$.

3. Simplify $\dfrac{9^{-1} x^{-5}}{3^{-3} x^{-2} y^0}$.

4. You are painting a rectangular canvas that is 21 inches wide and 27 inches long. Your friend is painting a rectangular canvas, where the width and length are each x inches shorter. When $x = 6$, what is the area of your friend's canvas?

In Exercises 7 and 8, find the product.

7. $(4x - 7)^2$

8. $(y + 3)(8 - 5y)$

9. Solve the system using any method.
   $2x + y = 3$
   $x - 3y = 5$

4.5 Self-Assessment

Use the scale to rate your understanding of the learning target and the success criteria.

| 1 I do not understand. | 2 I can do it with help. | 3 I can do it on my own. | 4 I can teach someone else. |

	Rating	Date
4.5 Dilations		
Learning Target: Understand dilations of figures.	1 2 3 4	
I can identify dilations.	1 2 3 4	
I can dilate figures.	1 2 3 4	
I can solve real-life problems involving scale factors and dilations.	1 2 3 4	

62 Geometry
Practice Workbook and Test Prep

Name_____ Date_____

4.6 Extra Practice

In Exercises 1 and 2, graph the polygon with the given vertices and its image after the similarity transformation.

1. $R(12, 8)$, $S(8, 0)$, $T(0, 4)$

   Dilation: $(x, y) \to \left(\frac{1}{4}x, \frac{1}{4}y\right)$

   Reflection: in the y-axis

2. $X(9, 6)$, $Y(3, 3)$, $Z(6, 3)$

   Rotation: 90° about the origin

   Dilation: $(x, y) \to \left(\frac{2}{3}x, \frac{2}{3}y\right)$

In Exercises 3 and 4, describe the similarity transformation that maps the preimage to the image.

3.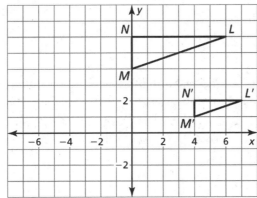

4.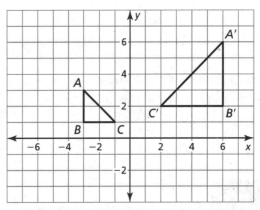

5. Prove that the figures are similar.

   Given Equilateral $\triangle GHI$ with side length a, equilateral $\triangle PQR$ with side length b

   Prove $\triangle GHI$ is similar to $\triangle PQR$.

   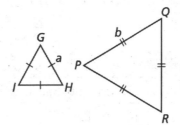

4.6 Review & Refresh

In Exercises 1–4, classify the angle.

1.

2.   44°

3. 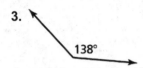 138°

4. ←——————→

5. Graph $\overline{PQ}$ with endpoints $P(-1, 4)$ and $Q(3, 1)$ and its image after the similarity transformation.
   Dilation: $(x, y) \rightarrow (2x, 2y)$
   Reflection: in the x-axis

6. Write an equation of the line passing through $(-2, 6)$ that is perpendicular to the line $y = -\frac{2}{3}x - 8$.

7. Solve $7x + 15 = 3x - 9$. Justify each step.

8. The linear function $j = 64 - 6g$ represents the amount of juice j (in fluid ounces) that you have left in a juice carton after pouring g glasses.

   a. Find the domain of the function. Is the domain discrete or continuous? Explain.

   b. Graph the function using its domain.

In Exercises 9 and 10, solve the inequality. Graph the solution.

9. $9 + 5x \leq 3x - 1$

10. $-3 < 2a + 3 < 15$

4.6 Self-Assessment

Use the scale to rate your understanding of the learning target and the success criteria.

| 1 I do not understand. | 2 I can do it with help. | 3 I can do it on my own. | 4 I can teach someone else. |

	Rating	Date
4.6 Similarity and Transformations		
Learning Target: Understand similarity transformations.	1 2 3 4	
I can perform similarity transformations.	1 2 3 4	
I can describe similarity transformations.	1 2 3 4	
I can prove that figures are similar.	1 2 3 4	

Name_____ Date_____

 Chapter Self-Assessment

Use the scale to rate your understanding of the learning target and the success criteria.

1 I do not understand. **2** I can do it with help. **3** I can do it on my own. **4** I can teach someone else.

	Rating	Date
Chapter 4 Transformations		
Learning Target: Understand transformations.	1 2 3 4	
I can identify transformations.	1 2 3 4	
I can perform translations, reflections, rotations, and dilations.	1 2 3 4	
I can describe congruence and similarity transformations.	1 2 3 4	
I can solve problems involving transformations.	1 2 3 4	
4.1 Translations		
Learning Target: Understand translations of figures.	1 2 3 4	
I can translate figures.	1 2 3 4	
I can write a translation rule for a given translation.	1 2 3 4	
I can explain what a rigid motion is.	1 2 3 4	
I can perform a composition of translations on a figure.	1 2 3 4	
4.2 Reflections		
Learning Target: Understand reflections of figures.	1 2 3 4	
I can reflect figures.	1 2 3 4	
I can perform compositions with reflections.	1 2 3 4	
I can identify line symmetry in polygons.	1 2 3 4	
4.3 Rotations		
Learning Target: Understand rotations of figures.	1 2 3 4	
I can rotate figures.	1 2 3 4	
I can perform compositions with rotations.	1 2 3 4	
I can identify rotational symmetry in polygons.	1 2 3 4	
4.4 Congruence and Transformations		
Learning Target: Understand congruence transformations.	1 2 3 4	
I can identify congruent figures.	1 2 3 4	
I can describe congruence transformations.	1 2 3 4	
I can use congruence transformations to solve problems.	1 2 3 4	

Name _____ Date _____

 Chapter Self-Assessment (continued)

	Rating	Date
4.5 Dilations		
Learning Target: Understand dilations of figures.	1 2 3 4	
I can identify dilations.	1 2 3 4	
I can dilate figures.	1 2 3 4	
I can solve real-life problems involving scale factors and dilations.	1 2 3 4	
4.6 Similarity and Transformations		
Learning Target: Understand similarity transformations.	1 2 3 4	
I can perform similarity transformations.	1 2 3 4	
I can describe similarity transformations.	1 2 3 4	
I can prove that figures are similar.	1 2 3 4	

Chapter 4 Test Prep

1. Select all transformations that do not result in mapping $(a, -a)$ to (a, a) when $a > 0$.

   Ⓐ reflection in the y-axis

   Ⓑ reflection in the x-axis

   Ⓒ translation $2a$ units up

   Ⓓ translation a units up, followed by a reflection in the line $y = a$

   Ⓔ translation a units up, followed by a reflection in the line $y = \dfrac{a}{2}$

2. What is the next number in the sequence $-3, -2, 1, 6, 13, \ldots$?

   _____

3. $\overrightarrow{BD}$ bisects $\angle ABC$. If $m\angle ABC = (6x + 58)°$, find $m\angle ABD$.

   Ⓐ 8°

   Ⓑ 16°

   Ⓒ 53°

   Ⓓ 106°

4. Use the dilation to find the value of y.

   Ⓐ 1.4

   Ⓑ 49

   Ⓒ 5

   Ⓓ 25

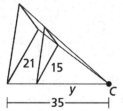

5. Which graph shows the image of $\triangle ABC$ after the glide reflection?
   Translation: $(x, y) \rightarrow (x + 2, y + 3)$
   Reflection: in the line $y = x$

   Ⓐ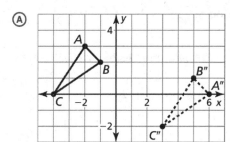

   Ⓑ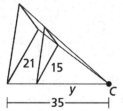

   Ⓒ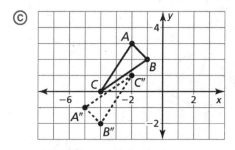

   Ⓓ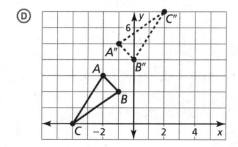

Chapter 4 Test Prep (continued)

6. What is the value of the expression $y - x$?

   Quadrilateral with angles $(7x + 40)°$, $(17y + 73)°$, $(68 - 6y)°$, $(121 - 4x)°$

   $y - x =$

7. The vector $\langle 2, 6 \rangle$ describes the translation of $M(3, 2w)$ to $M'(4x - 2, 7)$ and $N(5y - 8, 9)$ to $N'\left(0, \frac{3}{2}z\right)$. What is the sum of w, x, y, and z?

   $w + x + y + z =$

8. Which of the following is a congruence transformation that maps the preimage to the image?

   Ⓐ reflection in the x-axis, followed by a rotation of $180°$ about the origin

   Ⓑ reflection in the line $y = x$, followed by a reflection in the x-axis

   Ⓒ rotation of $270°$ about the origin, followed by a reflection in the y-axis

   Ⓓ reflection in the line $y = -x$, followed by a rotation of $180°$ about the origin

9. What are the vertices of the image of $\triangle XYZ$ after a reflection in the line $y = -x$, followed by a rotation of $270°$ about the origin?

   Ⓐ $X''(-1, -5)$, $Y''(1, 1)$, $Z''(3, 0)$

   Ⓑ $X''(-1, 5)$, $Y''(1, -1)$, $Z''(3, 0)$

   Ⓒ $X''(1, 5)$, $Y''(-1, -1)$, $Z''(-3, 0)$

   Ⓓ $X''(1, -5)$, $Y''(-1, 1)$, $Z''(-3, 0)$

Name_____ Date_____

Chapter 4 Test Prep (continued)

10. What is the area of the triangle?

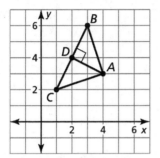

11. $AD = 29$. What is BC?

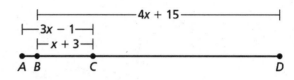

Ⓐ 3   Ⓑ 6
Ⓒ 8   Ⓓ 21

Ⓐ $\dfrac{\sqrt{5}}{2}$ square units

Ⓑ $\sqrt{5}$ square units

Ⓒ 5 square units

Ⓓ 10 square units

12. If the distance between m and n is 3.8 inches, what is the length of $\overline{RR''}$?

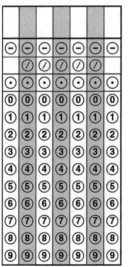

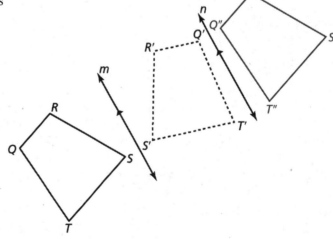

13. Write a rule for the translation of $\triangle ABC$ to $\triangle A'B'C'$.

_____

Copyright © Big Ideas Learning, LLC
All rights reserved.

Geometry
Practice Workbook and Test Prep   69

Chapter 4 Test Prep (continued)

14. What are the values of x that make $m \parallel n$?

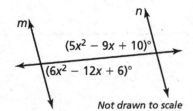
Not drawn to scale

Ⓐ $x = -1$ and $x = 4$

Ⓑ $x = -3$ and $x = 5$

Ⓒ $x = 24$ and $x = 54$

Ⓓ $x = 82$ and $x = 90$

15. Which of the following statements can you conclude from the diagram?

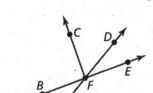

Ⓐ $\angle CFD$ and $\angle DFE$ are complementary.

Ⓑ $\angle CFD \cong \angle BFA$

Ⓒ $\overrightarrow{FC}$ bisects $\overline{BE}$.

Ⓓ $\angle BFA \cong \angle DFE$

16. What are the vertices of the image of $\triangle DEF$ after a dilation with scale factor 2, followed by a translation 2 units left and 4 units up?

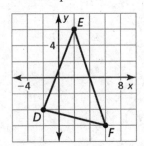

Ⓐ $D''(-3, 2), E''(-1, 7), F''(1, 1)$

Ⓑ $D''(-6, -4), E''(2, 16), F''(10, -8)$

Ⓒ $D''(-2, 0), E''(0, 5), F''(2, -1)$

Ⓓ $D''(-8, 0), E''(0, 20), F(8, -4)$

17. What is the distance from $A(-6, 1)$ to the line $3x - 4y = 16$?

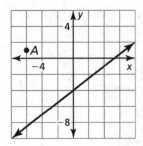

Ⓐ 0.5 unit

Ⓑ 7.6 units

Ⓒ 27.1 units

Ⓓ 57.8 units

18. Select all of the transformations that are equivalent to a rotation of 90° about the origin.

Ⓐ 270° clockwise about the origin

Ⓑ 90° clockwise about the origin

Ⓒ reflection in the line $y = -x$, followed by a reflection in the x-axis

Ⓓ reflection in the line $y = x$, followed by a reflection in the y-axis

Ⓔ reflection in the line $y = x$, followed by a reflection in the line $y = -x$

Name_____ Date_____

5.1 Extra Practice

In Exercises 1–3, classify the triangle by its sides and by measuring its angles.

1.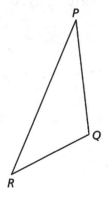

2.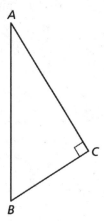

3.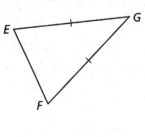

4. △ABC has vertices A(6, 6), B(9, 3), and C(2, 2). Classify the triangle by its sides. Then determine whether it is a right triangle.

In Exercises 5 and 6, find the measure of the exterior angle.

5.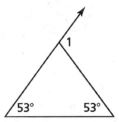

6.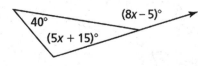

7. In a right triangle, the measure of one acute angle is twice the sum of the measure of the other acute angle and 30. Find the measure of each acute angle.

8. Your friend claims that the measure of an exterior angle of a triangle can never be acute because it is the sum of the measures of the two nonadjacent interior angles. Is your friend correct? Explain.

9. The figure shows the measures of angles of a roof truss. Find the measure of ∠1, the angle between an eave and a horizontal support beam.

Name_____   Date_____

5.1 Review & Refresh

1. Determine whether the triangles with vertices $D(-3, 5)$, $E(-1, 2)$, $F(-4, 4)$ and $G(-4, 1)$, $H(-2, -2)$, $I(-5, 0)$ are congruent. Use transformations to explain your reasoning.

2. Find the measure of each acute angle in a right triangle in which the measure of one acute angle is 4 times the measure of the other acute angle.

In Exercises 3 and 4, solve the equation.

3. $|11 - 4y| = 5$

4. $3t - 8 = -t$

5. Find the scale factor of the dilation. Then tell whether the dilation is a *reduction* or an *enlargement*.

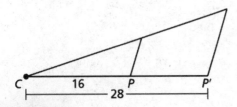

6. Determine whether $y = 0.7(1.3)^t$ represents *exponential growth* or *exponential decay*. Identify the percent rate of change.

7. Describe a similarity transformation that maps $\triangle LMN$ to $\triangle XYZ$.

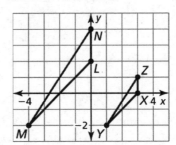

In Exercises 8 and 9, use the diagram.

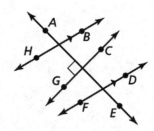

8. Name a pair of parallel lines.

9. Name a pair of perpendicular lines.

5.1 Self-Assessment

Use the scale to rate your understanding of the learning target and the success criteria.

| 1 I do not understand. | 2 I can do it with help. | 3 I can do it on my own. | 4 I can teach someone else. |

	Rating	Date
5.1 Angles of Triangles		
Learning Target: Prove and use theorems about angles of triangles.	1 2 3 4	
I can classify triangles by sides and by angles.	1 2 3 4	
I can prove theorems about angles of triangles.	1 2 3 4	
I can find interior and exterior angle measures of triangles.	1 2 3 4	

5.2 Extra Practice

In Exercises 1 and 2, identify all pairs of congruent corresponding parts. Then write another congruence statement for the polygons.

1. △PQR ≅ △STU

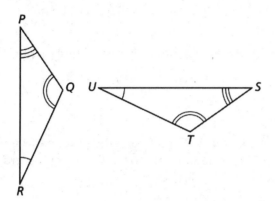

2. ABCD ≅ EFGH

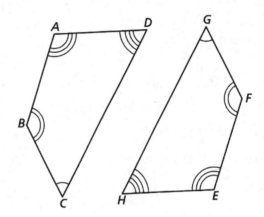

In Exercises 3 and 4, find the values of x and y.

3. △XYZ ≅ △RST

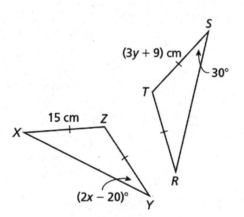

4. ABCD ≅ EFGH

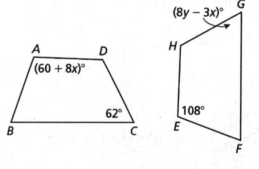

5. Show that the polygons are congruent. Explain your reasoning.

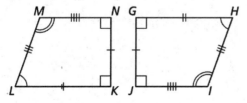

6. Find $m\angle 1$.

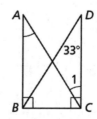

Name _____ Date _____

5.2 Review & Refresh

1. Find the measure of the exterior angle.

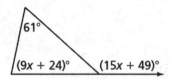

2. Graph △FGH with vertices $F(-4, 2)$, $G(2, 0)$, and $H(0, -4)$ and its image after the similarity transformation.

   Rotation: 90° about the origin

   Dilation: $(x, y) \rightarrow (\frac{3}{2}x, \frac{3}{2}y)$

4. You design a logo for your chemistry club. The logo is 2 inches by 3 inches. You decide to dilate the logo to 5 inches by 7.5 inches. What is the scale factor of this dilation?

In Exercises 5 and 6, factor the polynomial.

5. $x^2 - 2x - 15$ 6. $5x^2 + 17x - 12$

In Exercises 7 and 8, use the graphs of f and g to describe the transformation from the graph of f to the graph of g.

7. $f(x) = |x|;\ g(x) = -|2x| + 3$

8. $f(x) = x^3;\ g(x) = \frac{1}{3}x^3 - 1$

3. Write a congruence statement for the triangles. Identify all pairs of congruent corresponding parts.

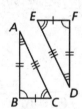

9. Write a piecewise function represented by the graph.

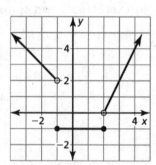

5.2 Self-Assessment

Use the scale to rate your understanding of the learning target and the success criteria.

| 1 I do not understand. | 2 I can do it with help. | 3 I can do it on my own. | 4 I can teach someone else. |

	Rating	Date
5.2 Congruent Polygons		
Learning Target: Understand congruence in terms of rigid motions.	1 2 3 4	
I can use rigid motions to show that two triangles are congruent.	1 2 3 4	
I can identify corresponding parts of congruent polygons.	1 2 3 4	
I can use congruent polygons to solve problems.	1 2 3 4	

74 Geometry
Practice Workbook and Test Prep

5.3 Extra Practice

1. Write a proof.

   Given $\overline{JN} \cong \overline{MN}, \overline{NK} \cong \overline{NL}$

   Prove $\triangle JNK \cong \triangle MNL$

   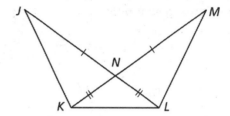

STATEMENTS	REASONS

In Exercises 2 and 3, use the given information to name two triangles that are congruent. Explain your reasoning.

2. $\angle EPF \cong \angle GPH$, and P is the center of the circle.

3. ABCDEF is a regular hexagon.

   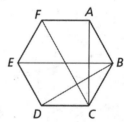

4. A quilt is made of triangles. You know $\overline{PS} \parallel \overline{QR}$ and $\overline{PS} \cong \overline{RQ}$. Use the SAS Congruence Theorem to show that $\triangle PQR \cong \triangle RSP$.

   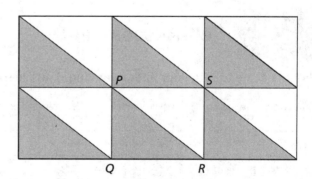

5. Your friend claims that the SAS Congruence Theorem will apply to a triangle and its image after the triangle has been translated, reflected, rotated, and dilated. Is your friend correct? Explain.

Name _____ Date _____

5.3 Review & Refresh

In Exercises 1 and 2, classify the triangle by its sides and by measuring its angles.

1.
2.

In Exercises 5 and 6, solve the inequality. Graph the solution.

5. $\dfrac{x}{2} + 8 \geq 5$

6. $|d - 3| < 9$

3. Graph $\triangle QRS$ with vertices $Q(-1, 2)$, $R(0, -1)$, and $S(-2, 1)$ and its image after the similarity transformation.

   Rotation: 180° about the origin
   Dilation: $(x, y) \to (3x, 3y)$

7. Find the values of x and y when $\triangle GHI \cong \triangle JKL$.

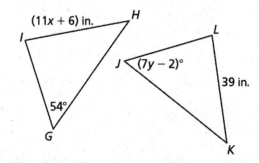

4. Decide whether enough information is given to prove that $\triangle PQR \cong \triangle STR$ by the SAS Congruence Theorem. Explain.

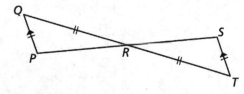

8. You want to determine the number of students in your school who own a cell phone. You survey 30 students at random. Twenty-four students own a cell phone, and six students do not. So, you conclude that 80% of the students in your school own a cell phone. Is your conclusion valid? Explain.

5.3 Self-Assessment

Use the scale to rate your understanding of the learning target and the success criteria.

| 1 I do not understand. | 2 I can do it with help. | 3 I can do it on my own. | 4 I can teach someone else. |

	Rating	Date
5.3 Proving Triangle Congruence by SAS		
Learning Target: Prove and use the Side-Angle-Side Congruence Theorem.	1 2 3 4	
I can use rigid motions to prove the SAS Congruence Theorem.	1 2 3 4	
I can use the SAS Congruence Theorem.	1 2 3 4	

76 Geometry
Practice Workbook and Test Prep

Name_____ Date_____

5.4 Extra Practice

In Exercises 1–4, complete the statement. State which theorem you used.

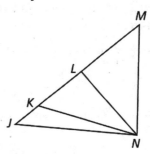

1. If $\overline{NJ} \cong \overline{NM}$, then $\angle$_____ $\cong \angle$_____.

2. If $\overline{LM} \cong \overline{LN}$, then $\angle$_____ $\cong \angle$_____.

3. If $\angle NKM \cong \angle NMK$, then _____ $\cong$ _____.

4. If $\angle LJN \cong \angle LNJ$, then _____ $\cong$ _____.

In Exercises 5 and 6, find the value of x.

5.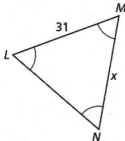

6.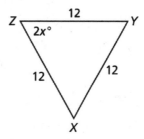

In Exercises 7 and 8, find the values of x and y.

7.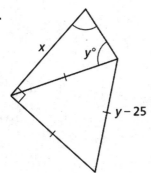

8.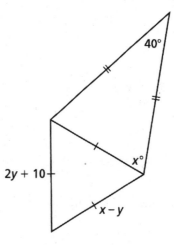

9. Explain why $\triangle ABC$ is isosceles.

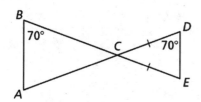

10. Can an isosceles triangle be an obtuse triangle? Explain.

Name _____ Date _____

5.4 Review & Refresh

In Exercises 1 and 2, use the given property to complete the statement.

1. Symmetric Property of Angle Congruence:

   If $\angle E \cong \angle H$, then _____ $\cong$ _____.

2. Transitive Property of Angle Congruence:

   If $\angle W \cong \angle I$ and _____ $\cong$ _____, then $\angle W \cong \angle T$.

3. Find $m\angle 1$.

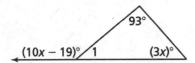

4. Graph $\overline{MN}$ with endpoints $M(2, 7)$ and $N(4, 1)$ and its image after the composition.
   Translation: $(x, y) \rightarrow (x + 2, y - 3)$
   Rotation: 90° about the origin

5. In the diagram, $ABCD \cong EFGH$. Find $m\angle F$ and GH.

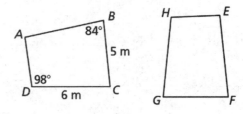

6. Find the distance from the point $(-6, -4)$ to the line $y = -3x + 8$.

7. Find the values of x and y.

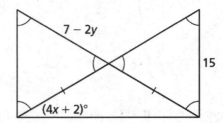

8. Find the mean, median, mode, range, and standard deviation of the data set.

   26, 32, 16, 22, 30, 19, 23

5.4 Self-Assessment

Use the scale to rate your understanding of the learning target and the success criteria.

| 1 I do not understand. | 2 I can do it with help. | 3 I can do it on my own. | 4 I can teach someone else. |

	Rating	Date
5.4 Equilateral and Isosceles Triangles		
Learning Target: Prove and use theorems about isosceles and equilateral triangles.	1 2 3 4	
I can prove and use theorems about isosceles triangles.	1 2 3 4	
I can prove and use theorems about equilateral triangles.	1 2 3 4	

5.5 Extra Practice

In Exercises 1 and 2, decide whether the congruence statement is true. Explain your reasoning.

1. △UVW ≅ △XYZ

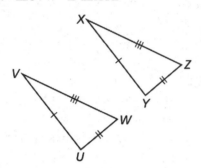

2. △KGH ≅ △HJK

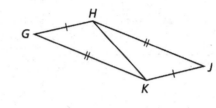

3. Write a proof.

   Given $\overline{MN} \cong \overline{PQ}$,

   ∠M and ∠P are right angles.

   Prove △MNQ ≅ △PQN

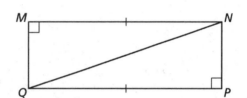

STATEMENTS	REASONS

4. Determine whether the figure is stable. Explain your reasoning.

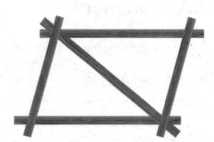

5. The figure shows a cut gem.

   a. What lengths can you measure to determine whether any two adjacent triangular faces of the gem are congruent?

   b. Assume that all of the triangular faces are congruent. What shape is the outline of the gem when viewed from above?

Name _____ Date _____

5.5 Review & Refresh

1. Are $\overrightarrow{UW}$ and $\overrightarrow{XZ}$ parallel? Explain your reasoning.

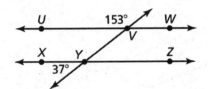

2. Find $m\angle 1$.

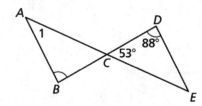

3. Write a linear function f with $f(1) = -1$ and $f(2) = 5$.

4. Find the values of x and y in the diagram of the roof.

5. Write a proof.
   Given E is the midpoint of $\overline{AC}$ and $\overline{BD}$.
   Prove $\triangle AEB \cong \triangle CED$

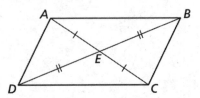

6. Graph $y > 2x - 5$ in a coordinate plane.

7. Decide whether the congruence statement $\triangle DOG \cong \triangle CAT$ is true. Explain your reasoning.

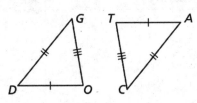

5.5 Self-Assessment

Use the scale to rate your understanding of the learning target and the success criteria.

| **1** I do not understand. | **2** I can do it with help. | **3** I can do it on my own. | **4** I can teach someone else. |

	Rating	Date
5.5 Proving Triangle Congruence by SSS		
Learning Target: Prove and use the Side-Side-Side Congruence Theorem.	1 2 3 4	
I can use rigid motions to prove the SSS Congruence Theorem.	1 2 3 4	
I can use the SSS Congruence Theorem.	1 2 3 4	
I can use the Hypotenuse-Leg Congruence Theorem.	1 2 3 4	

5.6 Extra Practice

In Exercises 1 and 2, decide whether enough information is given to prove that the triangles are congruent. If so, state the theorem you can use.

1. △GHK, △JKH

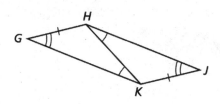

2. △ABC, △DEC

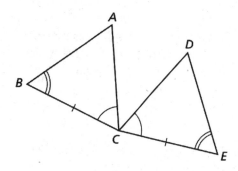

In Exercises 3 and 4, decide whether you can use the given information to prove that △LMN ≅ △PQR. Explain your reasoning.

3. ∠M ≅ ∠Q, ∠N ≅ ∠R, $\overline{NL} ≅ \overline{RP}$

4. ∠L ≅ ∠R, ∠M ≅ ∠Q, $\overline{LM} ≅ \overline{PQ}$

5. Prove that the triangles are congruent using the ASA Congruence Theorem.

   Given $\overline{AC}$ bisects ∠DAB and ∠DCB.

   Prove △ABC ≅ △ADC

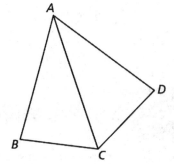

STATEMENTS	REASONS

6. Use the information given in the figure and the triangle congruence theorems to determine which pairs of triangles you can prove are congruent. Show your steps. Are there any pairs of triangles that cannot be proven congruent? Explain.

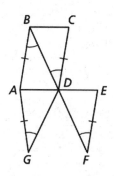

Name _____ Date _____

5.6 Review & Refresh

1. Find the coordinates of the midpoint of the line segment with endpoints $C(-5, 2)$ and $D(3, -6)$.

2. You know that a pair of triangles has two pairs of congruent corresponding sides. What other information do you need to show that the triangles are congruent?

5. You are using a microscope that shows the image of an object that is 10 times the object's actual size. Determine the length of the image of the bacteria seen through the microscope.

⊢— 0.4 mm —⊣

In Exercises 3 and 4, complete the statement. State which theorem you use.

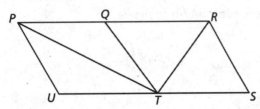

3. If $\angle QPT \cong \angle QTP$, then _____ ≅ _____.

4. If $\overline{QT} \cong \overline{QR}$, then $\angle$_____ ≅ $\angle$_____.

In Exercises 6 and 7, decide whether enough information is given to prove that the triangles are congruent. If so, state the theorem you can use.

6. △XYW, △ZYW   7. △ABC, △CDB

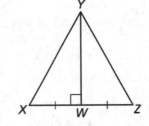 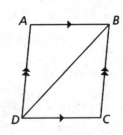

5.6 Self-Assessment

Use the scale to rate your understanding of the learning target and the success criteria.

| 1 I do not understand. | 2 I can do it with help. | 3 I can do it on my own. | 4 I can teach someone else. |

	Rating	Date
5.6 Proving Triangle Congruence by ASA and AAS		
Learning Target: Prove and use the Angle-Side-Angle Congruence Theorem and the Angle-Angle-Side Congruence Theorem.	1 2 3 4	
I can use rigid motions to prove the ASA Congruence Theorem.	1 2 3 4	
I can prove the AAS Congruence Theorem.	1 2 3 4	
I can use the ASA and AAS Congruence Theorems.	1 2 3 4	

Name_____ Date_____

5.7 Extra Practice

In Exercises 1 and 2, explain how to prove that the statement is true.

1. $\overline{UV} \cong \overline{XV}$

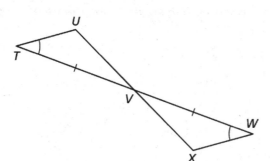

2. $\angle JLK \cong \angle MLN$

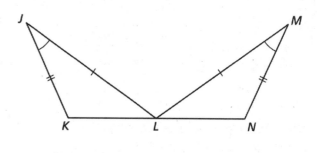

In Exercises 3 and 4, write a plan to prove that $\angle 1 \cong \angle 2$.

3.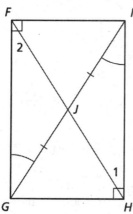

4.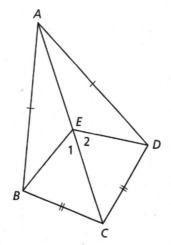

5. Write a proof to verify that the construction is valid.

   Ray bisects an angle

   Plan for Proof Show that $\triangle ABD \cong \triangle ACD$ by the SSS Congruence Theorem. Use corresponding parts of congruent triangles to show that $\angle BAD \cong \angle CAD$.

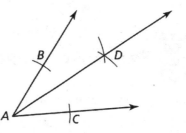

STATEMENTS	REASONS

Name _____ Date _____

5.7 Review & Refresh

1. Explain how you can prove that $\angle B \cong \angle D$ using a paragraph proof.

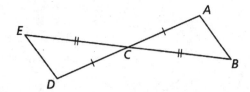

2. Find the perimeter of the polygon with vertices $J(-1, 1)$, $K(-3, 3)$, $L(-1, 5)$, and $M(1, 3)$.

3. Find the value of x.

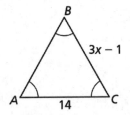

4. Simplify $(4m - 7) - (2 - 3m)$.

In Exercises 5 and 6, decide whether enough information is given to prove that the triangles are congruent. If so, state the theorem you can use.

5. $\triangle PQR$, $\triangle STU$

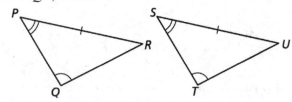

6. $\triangle WXY$, $\triangle YZW$

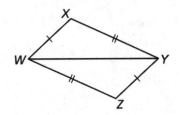

7. There are 2750 people in your town. The population of the town is increasing at an annual rate of 3.5%. Write and graph an exponential function to model the population of your town as a function of the number of years.

5.7 Self-Assessment

Use the scale to rate your understanding of the learning target and the success criteria.

| 1 | I do not understand. | 2 | I can do it with help. | 3 | I can do it on my own. | 4 | I can teach someone else. |

	Rating	Date
5.7 Using Congruent Triangles		
Learning Target: Use congruent triangles in proofs and to measure distances.	1 2 3 4	
I can use congruent triangles to prove statements.	1 2 3 4	
I can use congruent triangles to solve real-life problems.	1 2 3 4	
I can use congruent triangles to prove constructions.	1 2 3 4	

Name_____ Date_____

5.8 Extra Practice

In Exercises 1 and 2, place the figure in a coordinate plane in a convenient way. Assign coordinates to each vertex. Explain the advantages of your placement.

1. an obtuse triangle with a height of 3 units and a base of 2 units

2. a rectangle with a length of $2w$ and a base of 2 units

3. Graph the triangle with vertices $A(0, 0)$, $B(3m, m)$, and $C(0, 3m)$. Find the length and the slope of each side of the triangle. Then find the coordinates of the midpoint of each side. Is the triangle a right triangle? isosceles? Explain. (Assume all variables are positive.)

4. Write a plan for the proof.

   Given Coordinates of vertices of $\triangle OPR$ and $\triangle QRP$

   Prove $\triangle OPR \cong \triangle QRP$

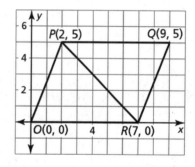

5. Write a coordinate proof.

   Given Coordinates of vertices of $\triangle OEF$ and $\triangle OGF$

   Prove $\triangle OEF \cong \triangle OGF$

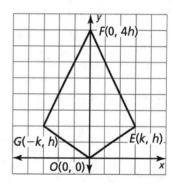

6. You design the front of a table that has three legs using a coordinate plane on a computer program. The coordinates of the vertices of the triangle formed by two legs and the floor are $A(0, 0)$, $B(11, 25)$, and $C(18, 0)$. One unit in the coordinate plane represents one inch. Prove that $\triangle ABC$ is a scalene triangle. Explain why the table may be unstable. Describe how to adjust point C to improve stability.

Copyright © Big Ideas Learning, LLC
All rights reserved.

Geometry
Practice Workbook and Test Prep

Name _____ Date _____

5.8 Review & Refresh

In Exercises 1 and 2, solve the equation. Justify each step.

1. $25 - 8x = 1$

2. $-4(x + 3) = 6 - x$

6. Write a proof.

   Given $\overline{AB} \cong \overline{DE}$, C is the center of the circle.

   Prove $\triangle ABC \cong \triangle DEC$

   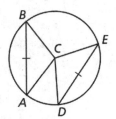

3. Factor the polynomial $21m^2 + 22m + 5$.

4. Explain how to prove that $\overline{SR} \cong \overline{UR}$.

   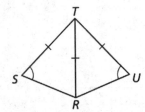

7. Place a rectangle with a length of $3w$ units in a coordinate plane in a convenient way for finding the length of the diagonal. Assign coordinates to each vertex.

5. Decide whether enough information is given to prove that $\triangle ABC$ and $\triangle GFH$ are congruent. If so, state the theorem you can use.

   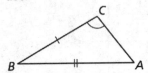 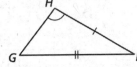

8. Find the value of x.

   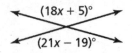

   $(18x + 5)°$
   $(21x - 19)°$

5.8 Self-Assessment

Use the scale to rate your understanding of the learning target and the success criteria.

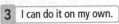

1 I do not understand. 2 I can do it with help. 3 I can do it on my own. 4 I can teach someone else.

	Rating	Date
5.8 Coordinate Proofs		
Learning Target: Use coordinates to write proofs.	1 2 3 4	
I can place figures in a coordinate plane.	1 2 3 4	
I can write plans for coordinate proofs.	1 2 3 4	
I can write coordinate proofs.	1 2 3 4	

Name_____ Date_____

Chapter 5 Chapter Self-Assessment

Use the scale to rate your understanding of the learning target and the success criteria.

1 I do not understand. **2** I can do it with help. **3** I can do it on my own. **4** I can teach someone else.

	Rating	Date
Chapter 5 Congruent Triangles		
Learning Target: Understand congruent triangles.	1 2 3 4	
I can classify triangles by sides and angles.	1 2 3 4	
I can solve problems involving congruent polygons.	1 2 3 4	
I can prove that triangles are congruent using different theorems.	1 2 3 4	
I can write a coordinate proof.	1 2 3 4	
5.1 Angles of Triangles		
Learning Target: Prove and use theorems about angles of triangles.	1 2 3 4	
I can classify triangles by sides and by angles.	1 2 3 4	
I can prove theorems about angles of triangles.	1 2 3 4	
I can find interior and exterior angle measures of triangles.	1 2 3 4	
5.2 Congruent Polygons		
Learning Target: Understand congruence in terms of rigid motions.	1 2 3 4	
I can use rigid motions to show that two triangles are congruent.	1 2 3 4	
I can identify corresponding parts of congruent polygons.	1 2 3 4	
I can use congruent polygons to solve problems.	1 2 3 4	
5.3 Proving Triangle Congruence by SAS		
Learning Target: Prove and use the Side-Angle-Side Congruence Theorem.	1 2 3 4	
I can use rigid motions to prove the SAS Congruence Theorem.	1 2 3 4	
I can use the SAS Congruence Theorem.	1 2 3 4	

Name _____ Date _____

Chapter 5 Chapter Self-Assessment (continued)

	Rating	Date
5.4 Equilateral and Isosceles Triangles		
Learning Target: Prove and use theorems about isosceles and equilateral triangles.	1 2 3 4	
I can prove and use theorems about isosceles triangles.	1 2 3 4	
I can prove and use theorems about equilateral triangles.	1 2 3 4	
5.5 Proving Triangle Congruence by SSS		
Learning Target: Prove and use the Side-Side-Side Congruence Theorem.	1 2 3 4	
I can use rigid motions to prove the SSS Congruence Theorem.	1 2 3 4	
I can use the SSS Congruence Theorem.	1 2 3 4	
I can use the Hypotenuse-Leg Congruence Theorem.	1 2 3 4	
5.6 Proving Triangle Congruence by ASA and AAS		
Learning Target: Prove and use the Angle-Side-Angle Congruence Theorem and the Angle-Angle-Side Congruence Theorem.	1 2 3 4	
I can use rigid motions to prove the ASA Congruence Theorem.	1 2 3 4	
I can prove the AAS Congruence Theorem.	1 2 3 4	
I can use the ASA and AAS Congruence Theorems.	1 2 3 4	
5.7 Using Congruent Triangles		
Learning Target: Use congruent triangles in proofs and to measure distances.	1 2 3 4	
I can use congruent triangles to prove statements.	1 2 3 4	
I can use congruent triangles to solve real-life problems.	1 2 3 4	
I can use congruent triangles to prove constructions.	1 2 3 4	
5.8 Coordinate Proofs		
Learning Target: Use coordinates to write proofs.	1 2 3 4	
I can place figures in a coordinate plane.	1 2 3 4	
I can write plans for coordinate proofs.	1 2 3 4	
I can write coordinate proofs.	1 2 3 4	

Chapter 5 Test Prep

1. Which of the following vertices form a right triangle?

   Ⓐ (3, −1), (4, 0), (6, −3)

   Ⓑ (8, 8), (5, 1), (3, 5)

   Ⓒ (0, 3), (7, 0), (1, 1)

   Ⓓ (−5, 2), (−4, 7), (−2, 4)

2. $ABCD \cong SRQT$. What is the value of x?

   Ⓐ 9

   Ⓑ 17

   Ⓒ 24

   Ⓓ 135

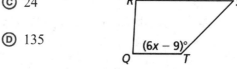

3. Find the area of a triangle with vertices $X(4, 3.5)$, $Y(4, 5)$, and $Z(7.25, 5)$.

   _____ square units

4. What is the distance from the point $(-3, 8)$ to the line $-4y = 8 - 3x$?

   _____ units

5. The vertex angle of isosceles $\triangle ABC$ is $\angle C$. What can you prove? Select all that apply.

   Ⓐ $\overline{AB} \cong \overline{BC}$

   Ⓑ $\angle A \cong \angle B$

   Ⓒ $\angle B \cong \angle C$

   Ⓓ $\overline{BC} \cong \overline{AC}$

   Ⓔ $\overline{AB} \cong \overline{AC}$

6. $\overline{XY}$ has endpoints $X(-3, 1)$ and $Y(4, -5)$. What is the midpoint of its image after a 90° rotation about the origin?

   _____

Chapter 5 Test Prep (continued)

7. Which of the following similarity transformations map △ABC to △A″B″C″? Select all that apply.

    Ⓐ translation 1 unit right and 2 units up, followed by a dilation with scale factor 2

    Ⓑ dilation with scale factor $\frac{1}{2}$, followed by a translation 1 unit left and 2 units down

    Ⓒ dilation with scale factor 2, followed by a translation 2 units right and 4 units up

    Ⓓ translation 2 units left and 4 units down, followed by a dilation with scale factor $\frac{1}{2}$

    Ⓔ dilation with scale factor 2, followed by a translation 2 units left and 4 units down

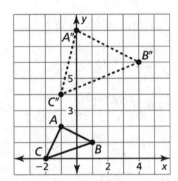

8. What is the length of the diagonal in the diagram?

    _____

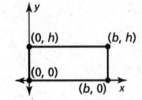

9. Two parallel lines are cut by a transversal. Which angle pairs are supplementary?

    Ⓐ corresponding angles

    Ⓑ alternate interior angles

    Ⓒ alternate exterior angles

    Ⓓ consecutive interior angles

10. In a right triangle, the measure of one acute angle is 2 times the difference of the other acute angle and 12. What is the measure of the smaller angle?

    Ⓐ 38°

    Ⓑ 52°

    Ⓒ 57°

    Ⓓ 34°

11. Which property of equality illustrates the statement "If $AC = 14$, then $BD + AC = BD + 14$?"

    Ⓐ Transitive Property of Equality

    Ⓑ Symmetric Property of Equality

    Ⓒ Division Property of Equality

    Ⓓ Substitution Property of Equality

12. What is the value of x?

    Ⓐ 6

    Ⓑ 9.75

    Ⓒ 17.25

    Ⓓ 30

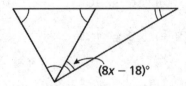

Chapter 5 Test Prep (continued)

13. Which of the following are corresponding angles?

Ⓐ ∠1 and ∠8

Ⓑ ∠2 and ∠10

Ⓒ ∠4 and ∠8

Ⓓ ∠2 and ∠5

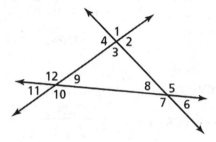

14. Which theorem can you use to prove that the triangles are congruent?

Ⓐ SSS Congruence Theorem

Ⓑ SAS Congruence Theorem

Ⓒ ASA Congruence Theorem

Ⓓ AAS Congruence Theorem

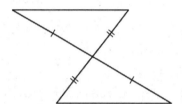

15. Find the value of x.

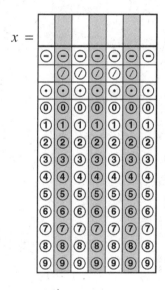

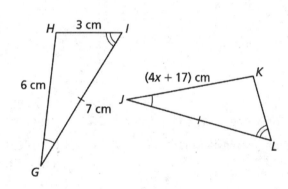

16. Two points lie in a plane. What can you conclude about the line containing the two points?

Ⓐ The line is parallel to the plane.

Ⓑ The line lies in the plane.

Ⓒ The line is perpendicular to the plane.

Ⓓ The line cannot exist.

Chapter 5 Test Prep (continued)

17. Which reason corresponds with the second statement in the proof, "$\overline{XZ} \cong \overline{ZX}$?"

Ⓐ Corresponding parts of congruent triangles are congruent.

Ⓑ Reflexive Property of Segment Congruence

Ⓒ Symmetric Property of Segment Congruence

Ⓓ Definition of congruent segments

Given $\overline{WX} \cong \overline{YZ}$, $\overline{XY} \cong \overline{ZW}$
Prove $\triangle WXZ \cong \triangle YZX$

STATEMENTS	REASONS
1. $\overline{WX} \cong \overline{YZ}$, $\overline{XY} \cong \overline{ZW}$	1. Given
2. $\overline{XZ} \cong \overline{ZX}$	2.
3. $\triangle WXZ \cong \triangle YZX$	3. SSS Congruence Theorem

18. What additional information do you need to prove that $\triangle PQR \cong \triangle SRQ$ by the ASA Congruence Theorem?

Ⓐ $\angle P \cong \angle S$

Ⓑ $\overline{PQ} \cong \overline{SR}$

Ⓒ $\angle QRS \cong \angle RQP$

Ⓓ $\overline{SQ} \cong \overline{PR}$

19. Which of the following is the composition rewritten as a single translation?

Translation: $(x, y) \rightarrow (x + 9, y - 8)$
Translation: $(x, y) \rightarrow (x - 3, y - 4)$

Ⓐ $(x, y) \rightarrow (x + 6, y - 12)$

Ⓑ $(x, y) \rightarrow (x + 12, y - 4)$

Ⓒ $(x, y) \rightarrow (x + 6, y - 4)$

Ⓓ $(x, y) \rightarrow (x + 12, y - 12)$

20. Rectangle $ABCD$ has vertices $A(0, 12)$, $B(6, 0)$, $C(0, -3)$, and $D(-6, 9)$. What is the area of the image after a dilation with a scale factor of $\frac{2}{3}$?

Ⓐ 90 square units

Ⓑ 40 square units

Ⓒ 202.5 square units

Ⓓ 60 square units

21. Complete the congruence statement.

$\triangle DEF \cong$ _____

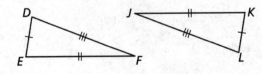

Name_____ Date_____

6.1 Extra Practice

In Exercises 1 and 2, tell whether the information in the figure allows you to conclude that point *P* lies on the perpendicular bisector of $\overline{LM}$. Explain your reasoning.

1.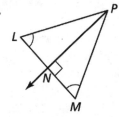

2.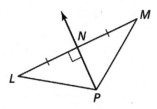

In Exercises 3–8, find the indicated measure. Explain your reasoning.

3. *AB*

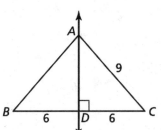

4. *EG*

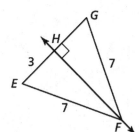

5. *SU*

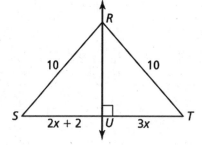

6. $m\angle CAB$

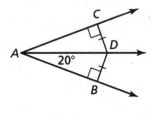

7. *CD*

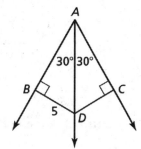

8. *BD*

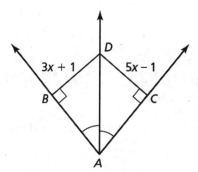

9. Write an equation of the perpendicular bisector of $\overline{AB}$ with endpoints $A(0, -2)$ and $B(2, 2)$.

10. Explain how you can use the perpendicular bisector of a segment to draw an isosceles triangle.

11. In a right triangle, is it possible for the bisector of the right angle to be the same line as the perpendicular bisector of the hypotenuse? Explain your reasoning. Draw a picture to support your answer.

Copyright © Big Ideas Learning, LLC
All rights reserved.

Name _____ Date _____

6.1 Review & Refresh

In Exercises 1 and 2, classify the triangle by its angles and sides.

1.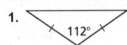

2.

3. A stair railing is designed as shown. Find $m\angle 2$. Explain your reasoning.

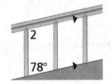

4. Use the given information to prove that $\triangle DAC \cong \triangle BCA$.

   Given $\angle ADE \cong \angle CBE$, $\overline{DE} \cong \overline{BE}$
   Prove $\triangle DAC \cong \triangle BCA$

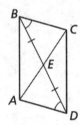

5. Find the product of $-3x^4$ and $2x^3 - 8x + 13$.

In Exercises 6 and 7, find the indicated measure. Explain your reasoning.

6. EG

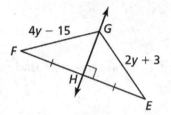

7. $m\angle LMN$

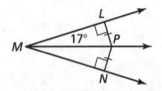

8. In $\triangle STU$ and $\triangle WXY$, $\overline{TU} \cong \overline{XY}$ and $\overline{SU} \cong \overline{WY}$. What is the third congruence statement that is needed to prove that $\triangle STU \cong \triangle WXY$ using the SSS Congruence Theorem? the SAS Congruence Theorem?

6.1 Self-Assessment

Use the scale to rate your understanding of the learning target and the success criteria.

| 1 I do not understand. | 2 I can do it with help. | 3 I can do it on my own. | 4 I can teach someone else. |

	Rating	Date
6.1 Perpendicular and Angle Bisectors		
Learning Target: Use theorems about perpendicular and angle bisectors.	1 2 3 4	
I can identify a perpendicular bisector and an angle bisector.	1 2 3 4	
I can use theorems about bisectors to find measures in figures.	1 2 3 4	
I can write equations of perpendicular bisectors.	1 2 3 4	

Name_____ Date_____

6.2 Extra Practice

In Exercises 1–3, find the indicated measure.

1. PA

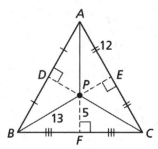

2. GE

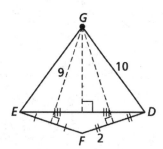

3. NF

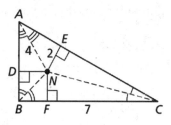

In Exercises 4 and 5, find the coordinates of the circumcenter of the triangle with the given vertices.

4. $A(-2, -2)$, $B(-2, 4)$, $C(6, 4)$

5. $D(3, 5)$, $E(3, 1)$, $F(9, 5)$

In Exercises 6–8, point N is the incenter of △ABC. Use the given information to find the indicated measure.

6. $ND = 2x - 5$
   $NE = -2x + 7$
   Find NF.

7. $NG = x - 1$
   $NH = 2x - 6$
   Find NJ.

8. $NK = x + 10$
   $NL = -2x + 1$
   Find NM.

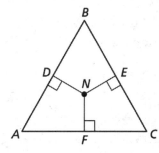

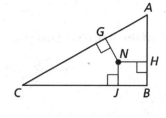

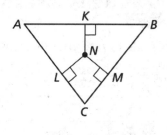

9. A cellular phone company is building a tower at an equal distance from each of three large apartment buildings. Explain how you can use the figure at the right to determine the location of the cell tower.

6.2 Review & Refresh

1. Determine whether $\triangle QRS$ and $\triangle TUV$ with the given vertices are congruent. Use transformations to explain your reasoning.

   $Q(-4, 5), R(2, 7), S(0, 2)$
   $T(-1, 3), U(5, 5), V(3, 0)$

2. Find $m\angle 1$. Then classify the triangle by its angles.

   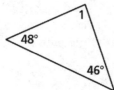

3. Explain how to prove that $\angle CDF \cong \angle EDF$.

   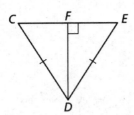

4. Factor $4h^6 - 64h^4$.

5. Find CD.

   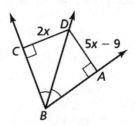

6. A triangle has vertices $X(-5, 1), Y(5, 3)$ and $Z(3, -7)$. Prove that $\triangle XYZ$ is isosceles.

7. The endpoints of $\overline{JK}$ are $J(-3, 6)$ and $K(-3, 0)$. Find the coordinates of the midpoint M. Then find JK.

8. Determine whether the table represents a *linear* or *nonlinear* function. Explain.

x	1	4	7	10
y	2	−2	−6	−10

6.2 Self-Assessment

Use the scale to rate your understanding of the learning target and the success criteria.

| 1 I do not understand. | 2 I can do it with help. | 3 I can do it on my own. | 4 I can teach someone else. |

	Rating	Date
6.2 Bisectors of Triangles		
Learning Target: Use bisectors of triangles.	1 2 3 4	
I can find the circumcenter and incenter of a triangle.	1 2 3 4	
I can circumscribe a circle about a triangle.	1 2 3 4	
I can inscribe a circle within a triangle.	1 2 3 4	
I can use points of concurrency to solve real-life problems.	1 2 3 4	

6.3 Extra Practice

1. Point P is the centroid of $\triangle LMN$, and $QN = 33$. Find PN and QP.

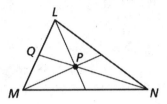

2. Point D is the centroid of $\triangle ABC$, and $DE = 7$. Find CD and CE.

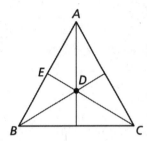

In Exercises 3 and 4, find the coordinates of the centroid of the triangle with the given vertices.

3. $A(-2, 1)$, $B(1, 8)$, $C(4, -1)$

4. $D(-5, 4)$, $E(-3, -2)$, $F(-1, 4)$

In Exercises 5 and 6, tell whether the orthocenter of the triangle with the given vertices is *inside*, *on*, or *outside* the triangle. Then find the coordinates of the orthocenter.

5. $X(3, 6)$, $Y(3, 0)$, $Z(11, 0)$

6. $P(3, 4)$, $Q(11, 4)$, $R(9, -2)$

7. To transport a triangular table, you remove the legs. You secure the glass top to the frame by looping a string from a hole in each vertex around the opposite side, then pulling it tight and tying it. At what point of concurrency do the three strings intersect? Explain your reasoning.

8. Is it impossible for the centroid and the orthocenter of a triangle to be the same point? Explain your reasoning.

Name_____ Date_____

6.3 Review & Refresh

1. Find $m\angle BDC$. Explain your reasoning.

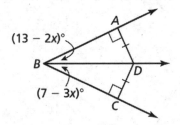

2. Find the coordinates of the circumcenter of the triangle with vertices $A(-2, 5)$, $B(4, 2)$, and $C(-2, -2)$.

3. Tell whether the orthocenter of $\triangle XYZ$ with vertices $X(-4, 2)$, $Y(6, 7)$, and $Z(-6, 16)$ is *inside*, *on*, or *outside* the triangle. Then find its coordinates.

4. $\overline{AB}$ has endpoints $A(5, 7)$ and $B(-1, 3)$. $\overline{CD}$ has endpoints $C(2, 1)$ and $D(4, -2)$. Is $\overline{AB} \parallel \overline{CD}$?

In Exercises 5 and 6, solve the system.

5. $7x = 11 - 2y$
   $2x + y = 1$

6. $4y - 3x = -7$
   $x - y = 2$

In Exercises 7 and 8, solve the equation.

7. $2n^2 + 50 = 0$

8. $x^2 - 15 = -4x$

9. You conduct a survey that asks 188 students in your school whether they plan to try out for the cross country team. Forty-nine of the students plan to try out, and 21 of those students are males. Sixty-seven of the females surveyed do not plan to try out. Organize the results in a two-way table. Include the marginal frequencies.

10. Write a plan for proof.

    Given $\overline{WY} \perp \overline{XZ}$
    Prove $\triangle WXY \cong \triangle WZY$

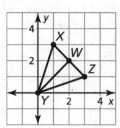

6.3 Self-Assessment

Use the scale to rate your understanding of the learning target and the success criteria.

| 1 I do not understand. | 2 I can do it with help. | 3 I can do it on my own. | 4 I can teach someone else. |

	Rating	Date
6.3 Medians and Altitudes of Triangles		
Learning Target: Use medians and altitudes of triangles.	1 2 3 4	
I can draw medians and altitudes of triangles.	1 2 3 4	
I can find the centroid of a triangle.	1 2 3 4	
I can find the orthocenter of a triangle.	1 2 3 4	

6.4 Extra Practice

In Exercises 1–3, $\overline{DE}$ is a midsegment of △ABC. Find the value of x.

1.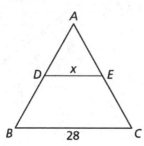
2.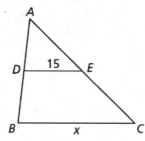
3.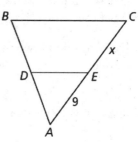

4. What is the perimeter of △DEF?

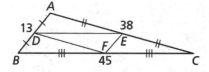

5. In the diagram, $\overline{DE}$ is a midsegment of △ABC, and $\overline{FG}$ is a midsegment of △ADE. Find FG.

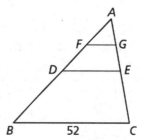

6. The area of △ABC is 48 square centimeters. $\overline{DE}$ is a midsegment of △ABC. What is the area of △ADE?

7. The diagram shows the front of a triangular woodshed. You are installing a shelf between two walls that is halfway up the 8-foot wall.

   a. How many feet should you measure from the ground along the slanted wall to find where to attach the opposite end of the shelf so that it will be level?

   b. How long is the shelf?

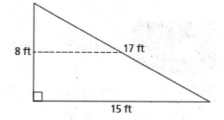

8. A building has the shape of a pyramid with a square base. The midsegment parallel to the ground of each triangular face of the pyramid has a length of 58 feet. Find the side length of the base of the pyramid.

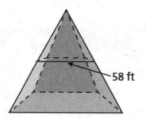

Name_____ Date_____

6.4 Review & Refresh

1. Find a counterexample to show that the conjecture is false.

   Conjecture The sum of two numbers is always positive.

2. Find *JK*. Explain your reasoning.

   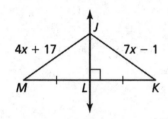

3. Find the coordinates of the centroid of $\triangle RST$ with vertices $R(3, 1)$, $S(7, 3)$, and $T(5, 5)$.

4. The incenter of $\triangle ABC$ is point *N*. $NQ = 4x + 3$ and $NS = 5x + 1$. Find *NR*.

   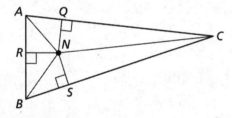

5. $\overline{MN}$ is a midsegment of $\triangle XYZ$. Find the values of *x* and *y*.

   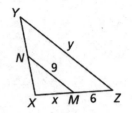

6. Write a piecewise function that represents the total cost *y* (in dollars) of renting a bicycle for *x* hours. Then determine the total cost for 435 minutes.

Bicycle Rentals	
Time	Price
up to 3 hours	$15
up to 6 hours	$25
up to 9 hours	$35
up to 12 hours	$45
$25 rental fee required	

   In Exercises 7 and 8, determine whether the equation represents a *linear* or *nonlinear* function. Explain.

7. $y = -1$

8. $y = 2x^2 - 3$

6.4 Self-Assessment

Use the scale to rate your understanding of the learning target and the success criteria.

| 1 I do not understand. | 2 I can do it with help. | 3 I can do it on my own. | 4 I can teach someone else. |

	Rating	Date
6.4 The Triangle Midsegment Theorem		
Learning Target: Find and use midsegments of triangles.	1 2 3 4	
I can use midsegments of triangles in the coordinate plane to solve problems.	1 2 3 4	
I can solve real-life problems involving midsegments.	1 2 3 4	

6.5 Extra Practice

In Exercises 1 and 2, write the first step in an indirect proof of the statement.

1. No number equals another number divided by zero.

2. The square root of 2 is not equal to the quotient of two integers.

In Exercises 3 and 4, determine which two statements contradict each other. Explain your reasoning.

3. A △LMN is equilateral.
   B LM ≠ MN
   C m∠L = m∠M

4. A △ABC is a right triangle.
   B ∠A is acute.
   C ∠C is obtuse.

5. List the angles of the triangle in order from smallest to largest.

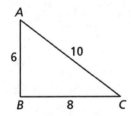

6. List the sides of the triangle in order from shortest to longest.

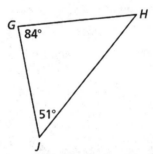

7. A triangle has one side of length 5 inches and another side of length 13 inches. Describe the possible lengths of the third side.

In Exercises 8–11, is it possible to construct a triangle with the given side lengths? If not, explain why not.

8. 3, 12, 17

9. 5, 21, 16

10. 8, 5, 7

11. 10, 3, 11

12. Describe the possible values of x.

Name_____ Date_____

6.5 Review & Refresh

1. △DEF has vertices D(2, 7), E(−4, 1), and F(4, 3). Find the coordinates of the vertices of the midsegment triangle of △DEF.

2. You have an old photograph that is 8 inches by 6 inches. You enlarge the photograph to 12 inches by 9 inches to hang on your wall. What is the scale factor of this dilation?

3. The incenter of △XYZ is point N. $NQ = 2x − 4$ and $NS = −3x + 11$. Find NR.

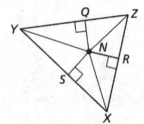

4. Tell whether the orthocenter of △QRS with vertices Q(5, 12), R(2, 3), and S(6, 7) is *inside*, *on*, or *outside* the triangle. Then find its coordinates.

5. Graph $\overline{XY}$ with endpoints X(3, −5) and Y(−2, 1) and its image after a reflection in the y-axis, followed by a rotation of 90° about the origin.

6. Decide whether enough information is given to prove that △ABE ≅ △DCE. If so, state the theorem you would use.

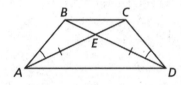

7. Find the value of x.

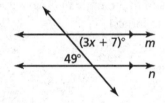

8. Solve $\frac{2}{5}x − \frac{4}{3}y = 2$ for y. Justify each step.

6.5 Self-Assessment

Use the scale to rate your understanding of the learning target and the success criteria.

| 1 | I do not understand. | 2 | I can do it with help. | 3 | I can do it on my own. | 4 | I can teach someone else. |

	Rating	Date
6.5 Indirect Proof and Inequalities in One Triangle		
Learning Target: Write indirect proofs and understand inequalities in a triangle.	1 2 3 4	
I can write indirect proofs.	1 2 3 4	
I can order the angles of a triangle given the side lengths.	1 2 3 4	
I can order the side lengths of a triangle given the angle measures.	1 2 3 4	
I can determine possible side lengths of triangles.	1 2 3 4	

Name_____ Date_____

6.6 Extra Practice

In Exercises 1–6, complete the statement with <, >, or =. Explain your reasoning.

1. BC_____EF

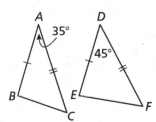

2. BC_____EF

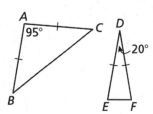

3. AB_____DC

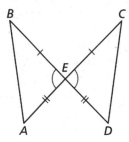

4. m∠A_____m∠D

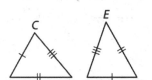

5. m∠A_____m∠D

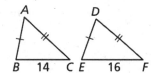

6. m∠1_____m∠2

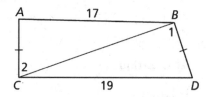

In Exercises 7 and 8, write a proof.

7. Given $\overline{XY} \cong \overline{ZY}$, WX > WZ

   Prove m∠WYX > m∠WYZ

   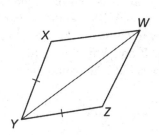

8. Given $\overline{AD} \cong \overline{CB}$, m∠DAC > m∠BCA

   Prove DC > BA

   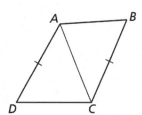

9. You loop a rubber band around the blade ends of a closed pair of scissors. Describe what happens to the rubber band as you open the scissors. How does this relate to the Hinge Theorem?

10. Starting from a point 10 miles north of Crow Valley, a crow turns 45° toward east and flies for 5 miles. Another crow, starting from a point 10 miles south of Crow Valley, flies 5 miles due west. Which crow is farther from Crow Valley? Explain.

6.6 Review & Refresh

In Exercises 1 and 2, find the value of x.

1.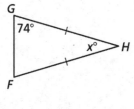

2. $\overline{DE}$ is a midsegment of $\triangle ABC$.
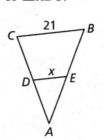

3. Graph quadrilateral WXYZ with vertices $W(2, 1)$, $X(5, -2)$, $Y(2, -5)$, and $Z(-1, -2)$ and its image after the similarity transformation.

   Translation: $(x, y) \rightarrow (x + 1, y + 2)$

   Dilation: $(x, y) \rightarrow \left(\frac{1}{3}x, \frac{1}{3}y\right)$

4. Which is longer, $\overline{AB}$ or $\overline{QR}$? Explain your reasoning.
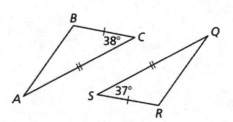

5. A city planner wants to build a gazebo. The planner sketches one possible location of the gazebo represented by point P.
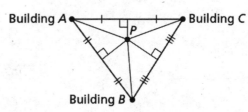

   a. Which point of concurrency did the planner use in the sketch as the location of the gazebo?

   b. The planner wants the gazebo to be equidistant from the sidewalks connecting the three buildings. Should the point of concurrency in the sketch be the location of the gazebo? If not, which point of concurrency should the planner use? Explain.

6.6 Self-Assessment

Use the scale to rate your understanding of the learning target and the success criteria.

| 1 I do not understand. | 2 I can do it with help. | 3 I can do it on my own. | 4 I can teach someone else. |

	Rating	Date
6.6 Inequalities in Two Triangles		
Learning Target: Understand inequalities in two triangles.	1 2 3 4	
I can explain the Hinge Theorem.	1 2 3 4	
I can compare measures in triangles.	1 2 3 4	
I can solve real-life problems using the Hinge Theorem.	1 2 3 4	

Name_____ Date_____

Chapter 6 Chapter Self-Assessment

Use the scale to rate your understanding of the learning target and the success criteria.

1 I do not understand. **2** I can do it with help. **3** I can do it on my own. **4** I can teach someone else.

	Rating	Date
Chapter 6 Relationships Within Triangles		
Learning Target: Understand relationships within triangles.	1 2 3 4	
I can identify and use perpendicular and angle bisectors of triangles.	1 2 3 4	
I can use medians and altitudes of triangles to solve problems.	1 2 3 4	
I can find distances using the Triangle Midsegment Theorem.	1 2 3 4	
I can compare measures within triangles and between two triangles.	1 2 3 4	
6.1 Perpendicular and Angle Bisectors		
Learning Target: Use theorems about perpendicular and angle bisectors.	1 2 3 4	
I can identify a perpendicular bisector and an angle bisector.	1 2 3 4	
I can use theorems about bisectors to find measures in figures.	1 2 3 4	
I can write equations of perpendicular bisectors.	1 2 3 4	
6.2 Bisectors of Triangles		
Learning Target: Use bisectors of triangles.	1 2 3 4	
I can find the circumcenter and incenter of a triangle.	1 2 3 4	
I can circumscribe a circle about a triangle.	1 2 3 4	
I can inscribe a circle within a triangle.	1 2 3 4	
I can use points of concurrency to solve real-life problems.	1 2 3 4	
6.3 Medians and Altitudes of Triangles		
Learning Target: Use medians and altitudes of triangles.	1 2 3 4	
I can draw medians and altitudes of triangles.	1 2 3 4	
I can find the centroid of a triangle.	1 2 3 4	
I can find the orthocenter of a triangle.	1 2 3 4	

Name _____ Date _____

Chapter Self-Assessment (continued)

	Rating	Date
6.4 The Triangle Midsegment Theorem		
Learning Target: Find and use midsegments of triangles.	1 2 3 4	
I can use midsegments of triangles in the coordinate plane to solve problems.	1 2 3 4	
I can solve real-life problems involving midsegments.	1 2 3 4	
6.5 Indirect Proof and Inequalities in One Triangle		
Learning Target: Write indirect proofs and understand inequalities in a triangle.	1 2 3 4	
I can write indirect proofs.	1 2 3 4	
I can order the angles of a triangle given the side lengths.	1 2 3 4	
I can order the side lengths of a triangle given the angle measures.	1 2 3 4	
I can determine possible side lengths of triangles.	1 2 3 4	
6.6 Inequalities in Two Triangles		
Learning Target: Understand inequalities in two triangles.	1 2 3 4	
I can explain the Hinge Theorem.	1 2 3 4	
I can compare measures in triangles.	1 2 3 4	
I can solve real-life problems using the Hinge Theorem.	1 2 3 4	

Name_____ Date_____

Chapter 6 Test Prep

1. What is the value of x?

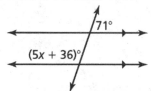

$x = $

2. You have a square digital photo that has a side length of 800 pixels. You enlarge the photo so its side length is 2100 pixels. What is the scale factor of this dilation?

$k = $

3. Point D is the centroid of $\triangle ABC$, and $DE = 6$. What is CD?

Ⓐ 4
Ⓑ 6
Ⓒ 12
Ⓓ 18

4. Which point lies on the perpendicular bisector of the segment with endpoints $A(-3, 6)$ and $B(5, 2)$?

Ⓐ $(4, 10)$
Ⓑ $(2, 2)$
Ⓒ $(3, 3)$
Ⓓ $(5, 6)$

5. Point N is the incenter of $\triangle ABC$. $ND = 4x + 5$ and $NE = 8x - 3$. What is NF?

Ⓐ 1
Ⓑ 2
Ⓒ 9
Ⓓ 13

6. Which of the following are not corresponding parts of the congruent triangles?

Ⓐ $\angle AEB \cong \angle CED$
Ⓑ $\angle B \cong \angle C$
Ⓒ $\overline{AB} \cong \overline{CD}$
Ⓓ $\angle A \cong \angle C$

Name _____ Date _____

Chapter 6 Test Prep (continued)

7. What is the perimeter of the triangle?

_____

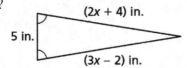

8. What is CE?

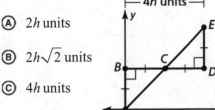

Ⓐ $2h$ units

Ⓑ $2h\sqrt{2}$ units

Ⓒ $4h$ units

Ⓓ $4h\sqrt{2}$ units

9. $\overline{DE}$ is a midsegment of $\triangle ABC$. What is BC?

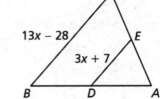

Ⓐ 6

Ⓑ 17.5

Ⓒ 35

Ⓓ 50

10. Three groups of hikers leave the same camp heading in different directions. Each group hikes 2.6 miles, then changes direction and hikes 3.2 miles. Group A starts due north and then turns 30° toward east. Group B starts due west and then turns 40° toward south. Group C starts due east and then turns 45° toward north. Which group is farther from camp?

Ⓐ Group A

Ⓑ Group B

Ⓒ Group C

Ⓓ All three groups are the same distance from camp.

11. $\triangle FGH \cong \triangle JKL$. What is the value of y?

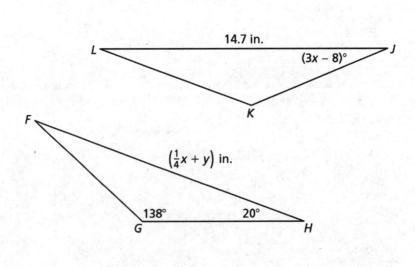

Chapter 6 Test Prep (continued)

12. Which of the following lists the angles of the triangle from smallest to largest?

Ⓐ ∠B, ∠C, ∠A
Ⓑ ∠A, ∠C, ∠B
Ⓒ ∠C, ∠B, ∠A
Ⓓ ∠B, ∠A, ∠C

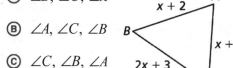

13. Which point of concurrency is point C?

Ⓐ circumcenter
Ⓑ incenter
Ⓒ centroid
Ⓓ orthocenter

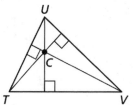

14. The line $y = -2x + 1$ is reflected in the line $y = 2$. Write the equation of the image.

_____

15. △XYZ has vertices $X(3, -2)$, $Y(5, 4)$, and $Z(-1, 6)$. Which of the following are the vertices of the midsegment triangle?

Ⓐ (1, 2)
Ⓑ (−1, −3)
Ⓒ (2, −4)
Ⓓ (4, 1)
Ⓔ (3, −1)
Ⓕ (2, 5)

16. In △DEF, which is a possible side length for $\overline{DF}$? Select all that apply.

Ⓐ 7.9
Ⓑ 8.2
Ⓒ 8.5
Ⓓ 8.9
Ⓔ 9.2

17. Which property does the statement "If $\overline{DE} \cong \overline{MN}$ and $\overline{MN} \cong \overline{XY}$, then $\overline{DE} \cong \overline{XY}$." illustrate?

Ⓐ Reflexive Property of Segment Congruence
Ⓑ Symmetric Property of Segment Congruence
Ⓒ Substitution Property of Equality
Ⓓ Transitive Property of Segment Congruence

Chapter 6 Test Prep (continued)

18. What is the perimeter of △QTS?

Ⓐ 13 units
Ⓑ 14 units
Ⓒ 16 units
Ⓓ 18 units

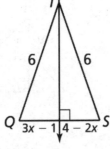

19. Which theorem can you use to prove that △LMN ≅ △QPR?

Ⓐ SSS Congruence Theorem
Ⓑ SAS Congruence Theorem
Ⓒ ASA Congruence Theorem
Ⓓ AAS Congruence Theorem

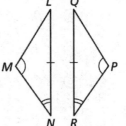

20. Which of the following statements is false?

Ⓐ The incenter of a triangle is equidistant from the sides of the triangle.

Ⓑ The circumcenter of a right triangle is on the triangle.

Ⓒ The incenter of an obtuse triangle is outside the triangle.

Ⓓ The circumcenter of a triangle is equidistance from the vertices of the triangle.

21. What is the angle of rotation that maps $\overline{CD}$ to $\overline{C''D''}$?

Ⓐ 26°
Ⓑ 52°
Ⓒ 78°
Ⓓ 104°

22. The midpoint of $\overline{RS}$ is $M(-7, 2)$. One endpoint is $S(-5, 6)$. What are the coordinates of endpoint R?

Ⓐ (−9, −2)
Ⓑ (−3, 10)
Ⓒ (−6, 4)
Ⓓ (−1, −2)

23. Which of the following is the inverse of the conditional statement?

Conditional statement If a polygon is an octagon, then it has eight sides.

Ⓐ If a polygon has eight sides, then it is an octagon.

Ⓑ If a polygon does not have eight sides, then it is not an octagon.

Ⓒ If a polygon is not an octagon, then it does not have eight sides.

Ⓓ A polygon is an octagon if and only if it has eight sides.

Name_____ Date_____

7.1 Extra Practice

In Exercises 1–3, find the sum of the measures of the interior angles of the indicated convex polygon.

1. octagon
2. 15-gon
3. 24-gon

In Exercises 4–6, the sum of the measures of the interior angles of a convex polygon is given. Classify the polygon by the number of sides.

4. 900°
5. 1620°
6. 2880°

In Exercises 7–10, find the value of x.

7.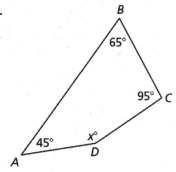

8.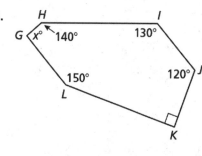

9.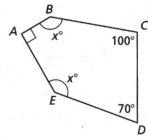

10.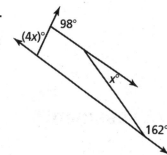

11. A pentagon has three interior angles that are congruent and two other interior angles that are supplementary to each other. Find the measure of each of the three congruent angles.

12. You are designing an amusement park ride with cars that spin in a circle around a center axis. The cars are located at the vertices of a regular polygon. The sum of the measures of the interior angles of the polygon is 6120°. If each car can hold four people, what is the maximum number of people who can be on the ride at one time?

Name _____ Date _____

7.1 Review & Refresh

In Exercises 1 and 2, find the value of x.

1.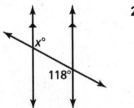

2.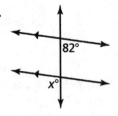

3. Which is greater, $m\angle 1$ or $m\angle 2$? Explain your reasoning.

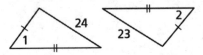

4. Describe the possible lengths of the third side of a triangle with side lengths of 14 feet and 6 feet.

5. Write an equation of the line that passes through $(8, -5)$ and is perpendicular to $y = -4x + 3$.

6. Determine whether the polygon has line symmetry. If so, draw the line(s) of symmetry and describe any reflections that map the figure onto itself.

7. $\overline{MN}$ is a midsegment of $\triangle PQR$. Find the value of x.

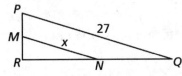

8. The sum of the measures of the interior angles of a convex polygon is 2340°. Classify the polygon by the number of sides.

9. Factor $x^2 - 5x - 66$.

10. Find the measure of the exterior angle.

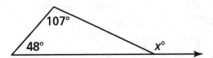

7.1 Self-Assessment

Use the scale to rate your understanding of the learning target and the success criteria.

1 I do not understand.   **2** I can do it with help.   **3** I can do it on my own.   **4** I can teach someone else.

	Rating	Date
7.1 Angles of Polygons		
Learning Target: Find angle measures of polygons.	1 2 3 4	
I can find the sum of the interior angle measures of a polygon.	1 2 3 4	
I can find interior angle measures of polygons.	1 2 3 4	
I can find exterior angle measures of polygons.	1 2 3 4	

112 Geometry
Practice Workbook and Test Prep

Name_____ Date_____

7.2 Extra Practice

In Exercises 1–3, find the value of each variable in the parallelogram.

1.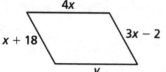
2.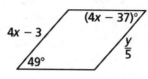
3.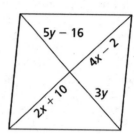

In Exercises 4–11, find the indicated measure in ▱MNOP. Explain your reasoning.

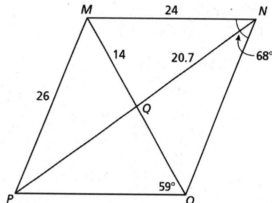

4. PO

5. OQ

6. NO

7. PQ

8. m∠PMN

9. m∠NOP

10. m∠OPM

11. m∠NMO

12. Write a two-column proof.

    Given: PQRS is a parallelogram.

    Prove: △PQT ≅ △RST

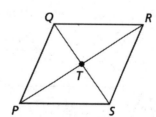

13. Three vertices of ▱WXYZ are W(–3, 4), Y(5, 3), and Z(3, 6). Find the coordinates of vertex X. Then find the coordinates of the intersection of the diagonals of ▱WXYZ.

Name _____ Date _____

7.2 Review & Refresh

1. List the sides of △ABC in order from shortest to longest.

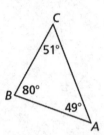

In Exercises 2–4, find the indicated measure in ▱QRST. Explain your reasoning.

2. QR

3. ∠S

4. ∠T

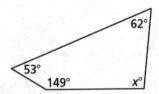

5. Find the value of x.

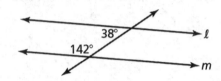

6. The coordinates of a point and its image after a reflection are shown. What is the line of reflection?

$$(-2, -9) \rightarrow (9, 2)$$

7. Decide whether there is enough information to prove that ℓ ∥ m. If so, state the theorem you can use.

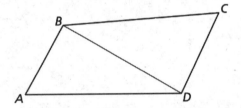

8. The hiking trail from A to B is shorter than the trail from C to D. The trail from A to D is the same length as the trail from C to B. What can you conclude about ∠ADB and ∠CBD? Explain your reasoning.

7.2 Self-Assessment

Use the scale to rate your understanding of the learning target and the success criteria.

| 1 I do not understand. | 2 I can do it with help. | 3 I can do it on my own. | 4 I can teach someone else. |

	Rating	Date
7.2 Properties of Parallelograms		
Learning Target: Prove and use properties of parallelograms.	1 2 3 4	
I can prove properties of parallelograms.	1 2 3 4	
I can use properties of parallelograms.	1 2 3 4	
I can solve problems involving parallelograms in the coordinate plane.	1 2 3 4	

7.3 Extra Practice

In Exercises 1–3, state which theorem you can use to show that the quadrilateral is a parallelogram.

1.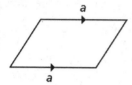
2.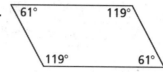
3.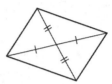

In Exercises 4–6, find the values of x and y that make the quadrilateral a parallelogram.

4.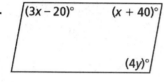
5.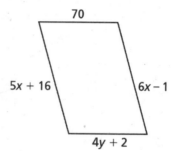
6.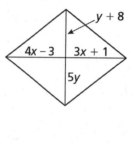

In Exercises 7 and 8, graph the quadrilateral with the given vertices in a coordinate plane. Then show that the quadrilateral is a parallelogram.

7. $J(-1, 2)$, $K(0, 4)$, $L(5, 4)$, $M(4, 2)$

8. $A(-2, -3)$, $B(1, -4)$, $C(6, 0)$, $D(3, 1)$

9. In the diagram of the handrail for a staircase, $m\angle CAB = 145°$ and $\overline{AB} \cong \overline{CD}$.

   a. Explain how to show that ABDC is a parallelogram.

   b. Describe how to prove that CDFE is a parallelogram.

   c. Can you prove that EFHG is a parallelogram? Explain.

   d. Find $m\angle ACD$, $m\angle DCE$, $m\angle CEF$, and $m\angle EFD$.

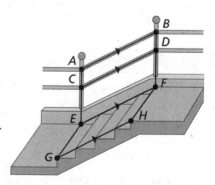

Name _____ Date _____

7.3 Review & Refresh

1. Solve the equation $4 - 2y = 5 - 6x$ for y. Justify each step.

2. Find the value of x.

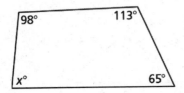

3. Find the distance between $X(-1, 5)$ and $Y(12, 2)$.

4. Three vertices of $\square ABCD$ are $A(-1, -4)$, $B(1, -1)$, and $C(-4, 1)$. Find the coordinates of the remaining vertex.

5. Graph $\triangle DEF$ with vertices $D(-1, 2)$, $E(1, 0)$, and $F(0, -1)$ and its image after a dilation with a scale factor of 2.

6. State which theorem you can use to show that the quadrilateral is a parallelogram.

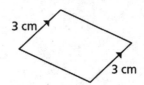

7. Place a rectangle with a length of 3ℓ units and a width of ℓ units in the coordinate plane. Find the length of the diagonal.

7.3 Self-Assessment

Use the scale to rate your understanding of the learning target and the success criteria.

| 1 I do not understand. | 2 I can do it with help. | 3 I can do it on my own. | 4 I can teach someone else. |

	Rating	Date
7.3 Proving That a Quadrilateral Is a Parallelogram		
Learning Target: Prove that a quadrilateral is a parallelogram.	1 2 3 4	
I can identify features of a parallelogram.	1 2 3 4	
I can prove that a quadrilateral is a parallelogram.	1 2 3 4	
I can find missing lengths that make a quadrilateral a parallelogram.	1 2 3 4	
I can show that a quadrilateral in the coordinate plane is a parallelogram.	1 2 3 4	

Name_____ Date_____

7.4 Extra Practice

1. For any rhombus *MNOP*, decide whether the statement $\overline{MO} \cong \overline{NP}$ is *always* or *sometimes* true. Draw a diagram and explain your reasoning.

2. For any rectangle *PQRS*, decide whether the statement $\angle PQS \cong \angle RSQ$ is *always* or *sometimes* true. Draw a diagram and explain your reasoning.

In Exercises 3–5, the diagonals of rhombus *ABCD* intersect at *E*. Given that $m\angle BCA = 44°$, $AB = 9$, and $AE = 7$, find the indicated measure.

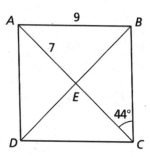

3. *BC*

4. *AC*

5. $m\angle ADC$

In Exercises 6–8, the diagonals of rectangle *EFGH* intersect at *I*. Given that $m\angle HFG = 31°$ and $EG = 17$, find the indicated measure.

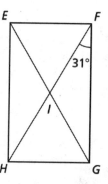

6. $m\angle FHG$

7. *HF*

8. $m\angle EFH$

In Exercises 9–11, the diagonals of square *LMNP* intersect at *K*. Given that $MK = \frac{1}{2}$, find the indicated measure.

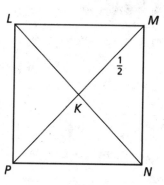

9. *PK*

10. $m\angle PKN$

11. $m\angle MNK$

In Exercises 12 and 13, decide whether ▱*JKLM* is a rectangle, a rhombus, or a square. Give all names that apply. Explain your reasoning.

12. $J(3, 2)$, $K(1, 1)$, $L(-1, 2)$, $M(1, 3)$

13. $J(-2, 5)$, $K(0, 7)$, $L(3, 4)$, $M(1, 2)$

Name _____ Date _____

7.4 Review & Refresh

In Exercises 1 and 2, use the graphs of f and g to describe the transformation from the graph of f to the graph of g.

1. $f(x) = 11x - 3$, $g(x) = f(x + 5)$

2. $f(x) = 15 - 8x$, $g(x) = f(3x)$

3. Rewrite the definition as a biconditional statement.

   Definition A *midsegment* of a triangle is a segment that connects the midpoints of two sides of the triangle.

In Exercises 4 and 5, solve the inequality. Graph the solution, if possible.

4. $|4m + 1| - 5 \leq -2$ 5. $9(t + 1) < 3(t + 9)$

6. Find the values of x and y in the parallelogram.

   53°, 4x − 9, 15, (18y − 1)°

7. Find the measure of each interior angle and each exterior angle of a regular 30-gon.

8. Find the perimeter and area of $\triangle XYZ$ with vertices $X(5, 1)$, $Y(-1, 1)$, and $Z(3, 2)$.

9. Decide whether you can use the given information $\angle D \cong \angle Q$, $\angle F \cong \angle S$, and $\overline{EF} \cong \overline{RS}$ to prove that $\triangle DEF \cong \triangle QRS$. Explain your reasoning.

10. Find the length of $\overline{AB}$. Explain your reasoning.

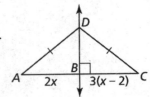

A, $2x$, B, $3(x - 2)$, C, D

7.4 Self-Assessment

Use the scale to rate your understanding of the learning target and the success criteria.

| 1 I do not understand. | 2 I can do it with help. | 3 I can do it on my own. | 4 I can teach someone else. |

	Rating	Date
7.4 Properties of Special Parallelograms		
Learning Target: Explain the properties of special parallelograms.	1 2 3 4	
I can identify special quadrilaterals.	1 2 3 4	
I can explain how special parallelograms are related.	1 2 3 4	
I can find missing measures of special parallelograms.	1 2 3 4	
I can identify special parallelograms in a coordinate plane.	1 2 3 4	

Name _____ Date _____

7.5 Extra Practice

1. Show that the quadrilateral with vertices $Q(0, 3)$, $R(0, 6)$, $S(-6, 0)$ and $T(-3, 0)$ is a trapezoid. Decide whether it is isosceles. Then find the length of its midsegment.

In Exercises 2 and 3, find $m\angle K$ and $m\angle L$.

2.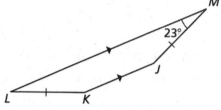

3.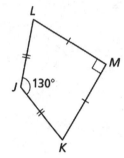

In Exercises 4 and 5, find CD.

4.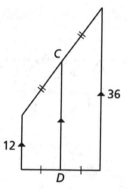

5.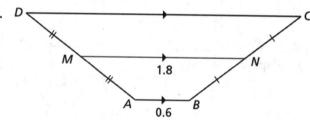

In Exercises 6 and 7, find the value of x.

6.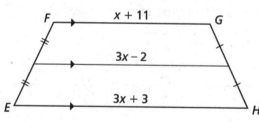

7.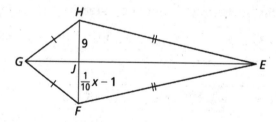

In Exercises 8 and 9, give the most specific name for the quadrilateral. Explain your reasoning.

8.

9.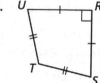

Name _____ Date _____

7.5 Review & Refresh

1. Decide whether enough information is given to prove that △RUT and △RUS are congruent using the HL Congruence Theorem.

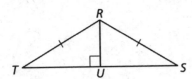

2. Find the distance from $(6, -1)$ to the line $y = x + 7$.

3. Classify the quadrilateral.

4. Find DB in $\square ABCD$. Explain your reasoning.

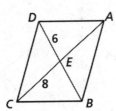

5. State which theorem you can use to show that the quadrilateral is a parallelogram.

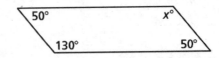

6. Graph $\overline{EF}$ with endpoints $E(2, 7)$ and $F(1, 4)$ and its image after a reflection in the y-axis, followed by a translation 3 units down.

7. Find the perimeter of the outer frame of the bridge.

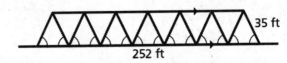

7.5 Self-Assessment

Use the scale to rate your understanding of the learning target and the success criteria.

| 1 I do not understand. | 2 I can do it with help. | 3 I can do it on my own. | 4 I can teach someone else. |

	Rating	Date
7.5 Properties of Trapezoids and Kites		
Learning Target: Use properties of trapezoids and kites to find measures.	1 2 3 4	
I can identify trapezoids and kites.	1 2 3 4	
I can use properties of trapezoids and kites to solve problems.	1 2 3 4	
I can find the length of the midsegment of a trapezoid.	1 2 3 4	
I can explain the hierarchy of quadrilaterals.	1 2 3 4	

120 Geometry
Practice Workbook and Test Prep

Name_____ Date_____

 Chapter Self-Assessment

Use the scale to rate your understanding of the learning target and the success criteria.

1 I do not understand. **2** I can do it with help. **3** I can do it on my own. **4** I can teach someone else.

	Rating	Date
Chapter 7 Quadrilaterals and Other Polygons		
Learning Target: Understand quadrilaterals and other polygons.	1　2　3　4	
I can find angles of polygons.	1　2　3　4	
I can describe properties of parallelograms.	1　2　3　4	
I can use properties of parallelograms.	1　2　3　4	
I can identify special quadrilaterals.	1　2　3　4	
7.1 Angles of Polygons		
Learning Target: Find angle measures of polygons.	1　2　3　4	
I can find the sum of the interior angle measures of a polygon.	1　2　3　4	
I can find interior angle measures of polygons.	1　2　3　4	
I can find exterior angle measures of polygons.	1　2　3　4	
7.2 Properties of Parallelograms		
Learning Target: Prove and use properties of parallelograms.	1　2　3　4	
I can prove properties of parallelograms.	1　2　3　4	
I can use properties of parallelograms.	1　2　3　4	
I can solve problems involving parallelograms in the coordinate plane.	1　2　3　4	
7.3 Proving That a Quadrilateral Is a Parallelogram		
Learning Target: Prove that a quadrilateral is a parallelogram.	1　2　3　4	
I can identify features of a parallelogram.	1　2　3　4	
I can prove that a quadrilateral is a parallelogram.	1　2　3　4	
I can find missing lengths that make a quadrilateral a parallelogram.	1　2　3　4	
I can show that a quadrilateral in the coordinate plane is a parallelogram.	1　2　3　4	

Name _____ Date _____

Chapter 7 Chapter Self-Assessment (continued)

	Rating	Date
7.4 Properties of Special Parallelograms		
Learning Target: Explain the properties of special parallelograms.	1 2 3 4	
I can identify special quadrilaterals.	1 2 3 4	
I can explain how special parallelograms are related.	1 2 3 4	
I can find missing measures of special parallelograms.	1 2 3 4	
I can identify special parallelograms in a coordinate plane.	1 2 3 4	
7.5 Properties of Trapezoids and Kites		
Learning Target: Use properties of trapezoids and kites to find measures.	1 2 3 4	
I can identify trapezoids and kites.	1 2 3 4	
I can use properties of trapezoids and kites to solve problems.	1 2 3 4	
I can find the length of the midsegment of a trapezoid.	1 2 3 4	
I can explain the hierarchy of quadrilaterals.	1 2 3 4	

Name_____ Date_____

Chapter 7 Test Prep

1. What is CD?

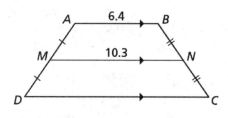

2. What is the measure of the exterior angle?

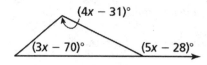

degrees

3. What is the value of x?

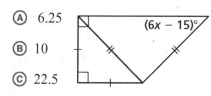

Ⓐ 6.25
Ⓑ 10
Ⓒ 22.5
Ⓓ 45

4. What is m∠LMN?

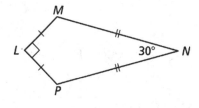

Ⓐ 60°
Ⓑ 90°
Ⓒ 120°
Ⓓ 150°

5. Which reason corresponds with the third statement in the proof, "∠ABC ≅ ∠DBE?"

Ⓐ Corresponding parts of congruent triangles are congruent.

Ⓑ Definition of congruent angles

Ⓒ Vertical Angles Congruence Theorem

Ⓓ Definition of angle bisector

Given $\overline{AC} \cong \overline{DE}$, ∠C ≅ ∠E, ∠A ≅ ∠D

Prove ∠ABC ≅ ∠DBE

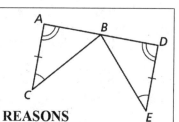

STATEMENTS	REASONS
1. $\overline{AC} \cong \overline{DE}$, ∠C ≅ ∠E, ∠A ≅ ∠D	1. Given
2. △ABC ≅ △DBE	2. ASA Congruence Theorem
3. ∠ABC ≅ ∠DBE	3.

Chapter 7 Test Prep (continued)

6. Which of the following statements is false?

- Ⓐ A square is a rhombus.
- Ⓑ A square is a parallelogram.
- Ⓒ A rectangle is a parallelogram.
- Ⓓ A parallelogram is a rhombus.

7. What is $m\angle F$?

- Ⓐ 89°
- Ⓑ 91°
- Ⓒ 96°
- Ⓓ 161°

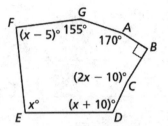

8. Three vertices of a parallelogram are $(-3, 1)$, $(-1, 4)$, and $(5, 1)$. Which of the following can be the fourth vertex of the parallelogram? Select all that apply.

- Ⓐ $(5, -1)$
- Ⓑ $(-1, -2)$
- Ⓒ $(3, -2)$
- Ⓓ $(3, 4)$
- Ⓔ $(-9, 4)$
- Ⓕ $(7, 4)$

9. Which of the following angle measures are possible exterior angle measures for regular polygons? Select all that apply.

- Ⓐ 8°
- Ⓑ 12°
- Ⓒ 54°
- Ⓓ 108°
- Ⓔ 120°
- Ⓕ 162°

10. What is the value of x?

- Ⓐ 3
- Ⓑ 4
- Ⓒ 6
- Ⓓ 8

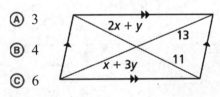

11. What is the 152nd term of the sequence A, G, T, C, A, G, T, C, A, G, T, C, …?

- Ⓐ A
- Ⓑ G
- Ⓒ T
- Ⓓ C

12. $\triangle JKL$ has vertices $J(-4, 5)$, $K(2, 3)$, and $L(0, 1)$. What is the perimeter of its midsegment triangle?

_____

Chapter 7 Test Prep (continued)

13. What is the most specific name for the quadrilateral with vertices (6, 8), (5, 6), (9, 7), and (10, 9)?

  Ⓐ parallelogram

  Ⓑ rhombus

  Ⓒ rectangle

  Ⓓ square

14. Which of the following would not provide enough information to prove that the quadrilateral is a parallelogram?

  Ⓐ $\overline{DE} \cong \overline{FG}, \overline{EF} \cong \overline{GD}$

  Ⓑ $\overline{EF} \cong \overline{GD}, \overline{EF} \parallel \overline{GD}$

  Ⓒ $\overline{DE} \parallel \overline{FG}, \overline{EF} \parallel \overline{GD}$

  Ⓓ $\overline{EF} \cong \overline{GD}, \overline{DE} \parallel \overline{FG}$

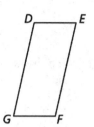

15. What can you conclude from the diagram?

  Ⓐ $EH = GH$

  Ⓑ $EH < GH$

  Ⓒ $EH > GH$

  Ⓓ No conclusion can be made.

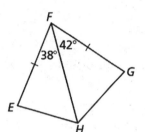

16. What is the distance between the point (3, 2) and its image after the composition?

  Translation: $(x, y) \rightarrow (x + 7, y - 1)$
  Translation: $(x, y) \rightarrow (x - 2, y + 13)$

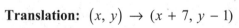 units

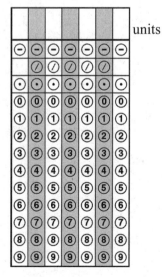

17. $\triangle ABC$ has vertices $A(-5, 8)$, $B(7, 8)$, and $C(7, 3)$. What is the difference of the perimeter of the image of $\triangle ABC$ and the perimeter of $\triangle ABC$ after the similarity transformation?

  Reflection: in the y-axis
  Dilation: $(x, y) \rightarrow (3x, 3y)$

  units

Chapter 7 Test Prep (continued)

18. What are the coordinates of the orthocenter of the triangle with vertices $W(2, 7)$, $X(3, 4)$, and $Y(6, 7)$?

   _____

19. What is the value of y?

    Ⓐ 4
    Ⓑ 10
    Ⓒ 28
    Ⓓ 30

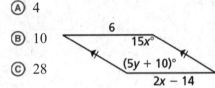

20. What is the value of x?

    Ⓐ 6.25
    Ⓑ 10.625
    Ⓒ 11.875
    Ⓓ 45

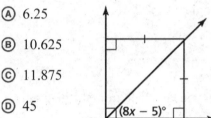

21. What can you conclude from the diagram?

    Ⓐ $a \perp k$
    Ⓑ $c \perp h$
    Ⓒ $a \parallel b$
    Ⓓ $a \parallel c$

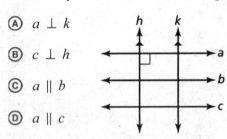

22. What is the value of y?

    Ⓐ 27
    Ⓑ 42
    Ⓒ 75
    Ⓓ 105

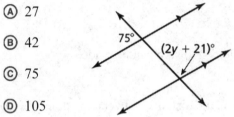

23. What rotations map the polygon onto itself? Select all that apply.

    Ⓐ 30°
    Ⓑ 60°
    Ⓒ 90°
    Ⓓ 120°
    Ⓔ 180°
    Ⓕ The polygon does not have rotational symmetry.

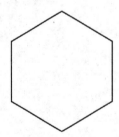

24. Which congruence statement is correct?

    Ⓐ $\triangle ABC \cong \triangle MNP$
    Ⓑ $\triangle ACB \cong \triangle MPN$
    Ⓒ $\triangle CAB \cong \triangle NMP$
    Ⓓ $\triangle BCA \cong \triangle PMN$

8.1 Extra Practice

In Exercises 1 and 2, the polygons are similar. Find the value of x.

1.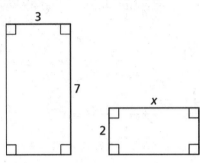

2.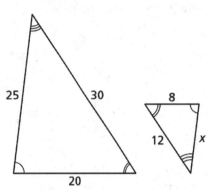

In Exercises 3–10, ABCDE ~ KLMNP.

3. Find the scale factor from ABCDE to KLMNP.

4. Find the scale factor from KLMNP to ABCDE.

5. List all pairs of congruent angles.

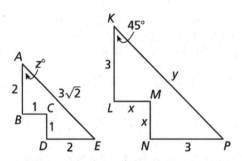

6. Write the ratios of the corresponding side lengths in a statement of proportionality.

7. Find the values of $x, y,$ and z.

8. Find the perimeter of each polygon.

9. Find the ratio of the perimeters of ABCDE to KLMNP.

10. Find the ratio of the areas of ABCDE to KLMNP.

11. Rhombus A is similar to rhombus B. Rhombus A has an area of 32 square feet. Rhombus B has an area of 98 square feet and a side length of 21 feet. Find the perimeter of rhombus A.

12. Tell whether two rectangles are *always*, *sometimes*, or *never* similar.

Name _____ Date _____

8.1 Review & Refresh

In Exercises 1 and 2, find the value of x.

1.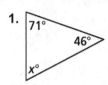

2.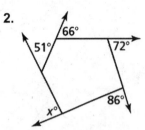

3. In rectangle WXYZ, $WY = 9x - 25$ and $XZ = 19 - 2x$. Find the lengths of the diagonals of WXYZ.

4. Find the values of x and y that make the quadrilateral a parallelogram.

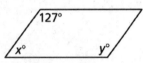

In Exercises 5 and 6, factor the polynomial.

5. $-5x^2 + 9x + 18$ 6. $x^3 - 2x^2 - 3x + 6$

7. The two triangles are similar. Find the value of x.

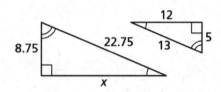

8. Tell whether the table of values represents a *linear*, an *exponential*, or a *quadratic* function.

x	−1	0	1	2	3
y	3	−1	3	15	35

In Exercises 9 and 10, solve the equation.

9. $3^{2(x+1)} = \left(\dfrac{1}{81}\right)^{x-5}$ 10. $4x^2 - 11 = 7$

11. The incenter of $\triangle ABC$ is point N, $NZ = 7x + 1$, and $NX = 2x + 6$. Find NY.

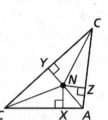

8.1 Self-Assessment

Use the scale to rate your understanding of the learning target and the success criteria.

1 I do not understand.   **2** I can do it with help.   **3** I can do it on my own.   **4** I can teach someone else.

	Rating	Date
8.1 Similar Polygons		
Learning Target: Understand the relationship between similar polygons.	1 2 3 4	
I can use similarity statements.	1 2 3 4	
I can find corresponding lengths in similar polygons.	1 2 3 4	
I can find perimeters and areas of similar polygons.	1 2 3 4	
I can decide whether polygons are similar.	1 2 3 4	

Name_____ Date_____

8.2 Extra Practice

1. Determine whether the triangles are similar. If they are, write a similarity statement. Explain your reasoning.

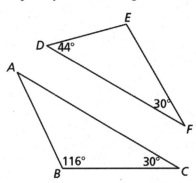

2. Show that the two triangles are similar and write a similarity statement.

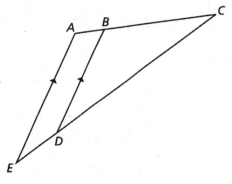

In Exercises 3–9, use the diagram to complete the statement.

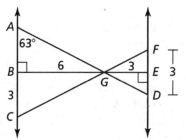

3. $m\angle AGB = $ ☐

4. $m\angle EGD = $ ☐

5. $m\angle BCG = $ ☐

6. $AB = $ ☐

7. $FE = $ ☐

8. $\triangle AGC \sim$ ☐

9. Write similarity statements for each triangle similar to $\triangle EFG$.

10. Use the following information to determine whether it is possible for $\triangle HJK$ and $\triangle RPQ$ to be similar. Explain your reasoning.

    $m\angle H = 100°, \ m\angle K = 46°, \ m\angle P = 44°, \text{ and } m\angle Q = 46°$

11. You can measure the width of the river using a surveying technique, as shown in the diagram. Find the width of the river, BC. Justify your answer.

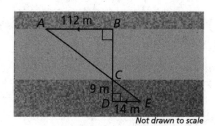

Copyright © Big Ideas Learning, LLC
All rights reserved.

Name _____ Date _____

8.2 Review & Refresh

1. Decide whether enough information is given to prove that the triangles are congruent. If so, state the theorem you can use.

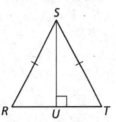

2. Determine whether the triangles are similar. If they are, write a similarity statement. Explain your reasoning.

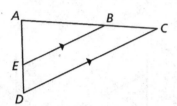

3. A bolt has a head shaped like a regular hexagon. Find the measure of each (a) interior angle and (b) exterior angle.

4. Find $m\angle N$.

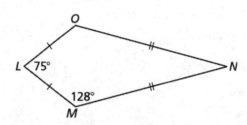

5. In the diagram, $QRST \sim WXYZ$. The area of $WXYZ$ is 64 square feet. Find the area of $QRST$.

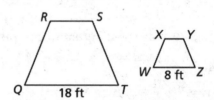

6. Decide whether ▱ABCD with vertices $A(-3, 4)$, $B(-1, 1)$, $C(2, 3)$, and $D(0, 6)$ is a *rectangle*, a *rhombus*, or a *square*. Give all names that apply.

8.2 Self-Assessment

Use the scale to rate your understanding of the learning target and the success criteria.

| 1 | I do not understand. | 2 | I can do it with help. | 3 | I can do it on my own. | 4 | I can teach someone else. |

	Rating	Date
8.2 Proving Triangle Similarity by AA		
Learning Target: Understand and use the Angle-Angle Similarity Theorem.	1 2 3 4	
I can use similarity transformations to prove the Angle-Angle Similarity Theorem.	1 2 3 4	
I can use angle measures of triangles to determine whether triangles are similar.	1 2 3 4	
I can prove triangle similarity using the Angle-Angle Similarity Theorem.	1 2 3 4	
I can solve real-life problems using similar triangles.	1 2 3 4	

8.3 Extra Practice

1. Determine whether △JKL or △RST is similar to △ABC.

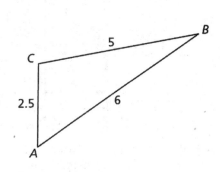
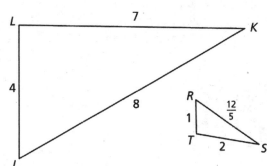

2. Find the value of x that makes △RST ~ △HGK.

3. Show that the triangles are similar and write a similarity statement. Explain your reasoning.

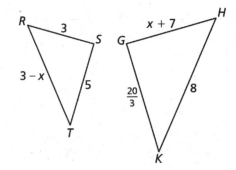
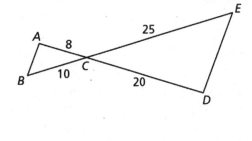

4. Use the following information to verify that △RST ~ △XYZ. Find the scale factor of △RST to △XYZ.

   △RST: RS = 12, ST = 15, TR = 24
   △XYZ: XY = 28, YZ = 35, ZX = 56

In Exercises 5 and 6, use △ABC.

5. The shortest side of a triangle similar to △ABC is 15 units long. Find the other side lengths of the triangle.

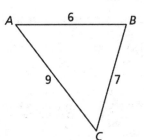

6. The longest side of a triangle similar to △ABC is 6 units long. Find the other side lengths of the triangle.

7. Determine whether △DEF with vertices D(2, 5), E(1, 4), and F(4, 3) is similar to △PQR with vertices P(1, 5), Q(−2, 2,), and R(7, −1). Explain your reasoning.

Name _____ Date _____

8.3 Review & Refresh

1. A xylophone consists of wooden bars. Each bar is parallel to the bar directly to the right. Explain why the longest bar is parallel to the shortest bar.

2. Show that △ABD and △CBA are similar. Then write a similarity statement.

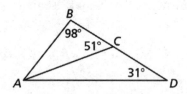

3. Find the coordinates of point P along the directed line segment AB with vertices A(−1, 1) and B(7, 3) so that the ratio of AP to PB is 1 to 3.

4. A triangle has side lengths of 7 meters and 16 meters. Describe the possible lengths of the third side.

5. Determine whether the graph represents a function. Explain.

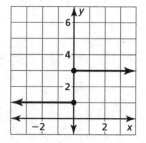

6. Write an equation of the perpendicular bisector of the segment with endpoints T(4, 9) and U(2, 3).

In Exercises 7 and 8, find $m\angle C$.

7.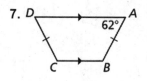

8.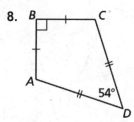

9. Decide whether enough information is given to prove that △JKM ≅ △LKM. If so, state the theorem you can use.

8.3 Self-Assessment

Use the scale to rate your understanding of the learning target and the success criteria.

| 1 I do not understand. | 2 I can do it with help. | 3 I can do it on my own. | 4 I can teach someone else. |

	Rating	Date
8.3 Proving Triangle Similarity by SSS and SAS		
Learning Target: Understand and use additional triangle similarity theorems.	1 2 3 4	
I can use the SSS and SAS similarity theorems to determine whether triangles are similar.	1 2 3 4	
I can use similar triangles to prove theorems about slopes of parallel and perpendicular lines.	1 2 3 4	

132 Geometry
Practice Workbook and Test Prep

Name_____ Date_____

8.4 Extra Practice

In Exercises 1 and 2, find the length of $\overline{AB}$.

1.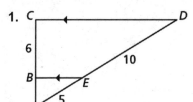

2.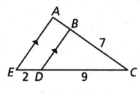

In Exercises 3 and 4, determine whether $\overline{AB} \parallel \overline{XY}$.

3.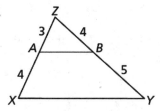

4.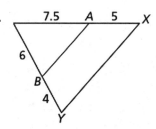

In Exercises 5–7, use the diagram to complete the proportion.

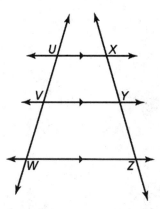

5. $\dfrac{UV}{UW} = \dfrac{XY}{\boxed{}}$

6. $\dfrac{XY}{YZ} = \dfrac{\boxed{}}{VW}$

7. $\dfrac{\boxed{}}{ZY} = \dfrac{WU}{WV}$

In Exercises 8 and 9, find the value of the variable.

8.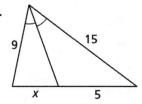

9.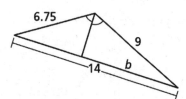

10. The diagram shows the skyline of a city. Find the distance between point E and point F for which $\overline{BE} \parallel \overline{CF}$. Explain your reasoning.

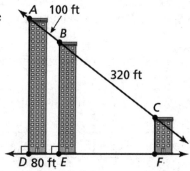

Name _____ Date _____

8.4 Review & Refresh

1. Graph $\triangle JKL$ with vertices $J(-1, 2)$, $K(0, 1)$, and $L(-2, -1)$ and its image after the similarity transformation.

   Dilation: $(x, y) \to (3x, 3y)$

   Reflection: in the x-axis

3. Find the value of x when $\triangle ABC \sim \triangle RST$.

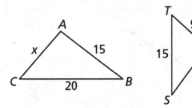

4. Solve the equation $A = \frac{1}{2}h(b_1 + b_2)$ for b_1.

5. Show that the triangles are similar. Write a similarity statement.

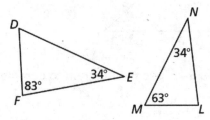

2. You are building a birdhouse with the triangular front shown. You want to make the entrance the same distance from each side of the triangular front. Determine the location of the entrance.

6. Find the value of x.

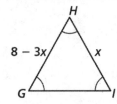

8.4 Self-Assessment

Use the scale to rate your understanding of the learning target and the success criteria.

| 1 I do not understand. | 2 I can do it with help. | 3 I can do it on my own. | 4 I can teach someone else. |

	Rating	Date
8.4 Proportionality Theorems		
Learning Target: Understand and use proportionality theorems.	1 2 3 4	
I can use proportionality theorems to find lengths in triangles.	1 2 3 4	
I can find lengths when two transversals intersect three parallel lines.	1 2 3 4	
I can find lengths when a ray bisects an angle of a triangle.	1 2 3 4	

134 Geometry
Practice Workbook and Test Prep

Name_____ Date_____

 Chapter Self-Assessment

Use the scale to rate your understanding of the learning target and the success criteria.

1 I do not understand. **2** I can do it with help. **3** I can do it on my own. **4** I can teach someone else.

	Rating	Date
Chapter 8 Similarity		
Learning Target: Understand similarity.	1 2 3 4	
I can identify corresponding parts of similar polygons.	1 2 3 4	
I can find and use scale factors in similar polygons.	1 2 3 4	
I can prove triangles are similar.	1 2 3 4	
I can use proportionality theorems to solve problems.	1 2 3 4	
8.1 Similar Polygons		
Learning Target: Understand the relationship between similar polygons.	1 2 3 4	
I can use similarity statements.	1 2 3 4	
I can find corresponding lengths in similar polygons.	1 2 3 4	
I can find perimeters and areas of similar polygons.	1 2 3 4	
I can decide whether polygons are similar.	1 2 3 4	
8.2 Proving Triangle Similarity by AA		
Learning Target: Understand and use the Angle-Angle Similarity Theorem.	1 2 3 4	
I can use similarity transformations to prove the Angle-Angle Similarity Theorem.	1 2 3 4	
I can use angle measures of triangles to determine whether triangles are similar.	1 2 3 4	
I can prove triangle similarity using the Angle-Angle Similarity Theorem.	1 2 3 4	
I can solve real-life problems using similar triangles.	1 2 3 4	
8.3 Proving Triangle Similarity by SSS and SAS		
Learning Target: Understand and use additional triangle similarity theorems.	1 2 3 4	
I can use the SSS and SAS similarity theorems to determine whether triangles are similar.	1 2 3 4	
I can use similar triangles to prove theorems about slopes of parallel and perpendicular lines.	1 2 3 4	

Name _____ Date _____

 Chapter Self-Assessment (continued)

	Rating	Date
8.4 Proportionality Theorems		
Learning Target: Understand and use proportionality theorems.	1 2 3 4	
I can use proportionality theorems to find lengths in triangles.	1 2 3 4	
I can find lengths when two transversals intersect three parallel lines.	1 2 3 4	
I can find lengths when a ray bisects an angle of a triangle.	1 2 3 4	

Chapter 8 Test Prep

1. Rectangle A is similar to rectangle B. Rectangle A has side lengths of 8 and 20. Rectangle B has a side length of 10. What are the possible values for the length of the other side of rectangle B? Select all that apply.

   Ⓐ 4

   Ⓑ 16

   Ⓒ 25

   Ⓓ 40

2. Which of the following statements is false?

   Ⓐ $AC < BC$

   Ⓑ $AB < AC + BC$

   Ⓒ $BC > AB$

   Ⓓ $\triangle ABC$ is scalene.

   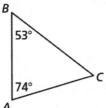

3. In the diagram, $\frac{AC}{DC} = \frac{BC}{EC}$. Which of the following statements is false?

   Ⓐ $\overline{AB} \parallel \overline{DE}$

   Ⓑ $\triangle ABC \sim \triangle DEC$

   Ⓒ $\angle A \cong \angle D$

   Ⓓ $\angle ABC \cong \angle EDC$

   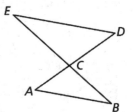

4. In ▱PQRS, the ratio of QR to RS is 3 to 2. The perimeter of ▱PQRS is 70 units. What is PS? _____ units

5. A tree casts a shadow that is 90 feet long. A person standing nearby who is 5 feet 6 inches casts a shadow that is 72 inches long. How tall is the tree? _____ feet

6. What are the coordinates of the circumcenter of the triangle with vertices $(-2, -1)$, $(-6, -1)$, and $(-6, 11)$.

Chapter 8 Test Prep (continued)

7. Which figure is stable?

Ⓐ    Ⓑ

Ⓒ    Ⓓ

8. Which name can be used to classify the quadrilateral? Select all that apply.

   Ⓐ parallelogram
   Ⓑ rectangle
   Ⓒ rhombus
   Ⓓ trapezoid

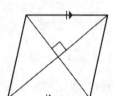

9. Which geometric figure illustrates the graph of $2x + 3 \leq 7$?

   Ⓐ point
   Ⓑ line
   Ⓒ line segment
   Ⓓ ray

10. What is the scale factor from $\triangle ABC$ to $\triangle DEF$?

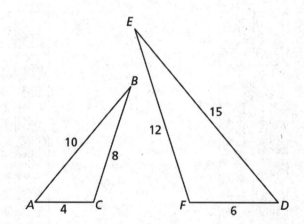

138 Geometry
Practice Workbook and Test Prep

Chapter 8 Test Prep (continued)

11. What is AC?

Ⓐ $5\frac{1}{4}$

Ⓑ 12

Ⓒ $12\frac{1}{4}$

Ⓓ $16\frac{1}{3}$

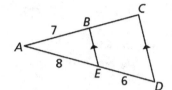

12. What is the value of z?

Ⓐ 1.8

Ⓑ 4

Ⓒ 5

Ⓓ 7.2

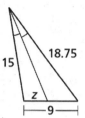

13. Figure X is similar to figure Y. Figure X has a perimeter of 22 inches and an area of 80 square inches. Figure Y has a perimeter of 55 inches. What is the area of figure Y?

Ⓐ 12.8 square inches

Ⓑ 32 square inches

Ⓒ 200 square inches

Ⓓ 500 square inches

14. What value of x makes $\triangle ABC \sim \triangle XYZ$?

Ⓐ 3

Ⓑ 4

Ⓒ 9

Ⓓ 10

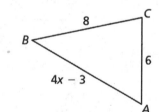

15. What is TU?

Ⓐ 8

Ⓑ 18

Ⓒ 20

Ⓓ 32

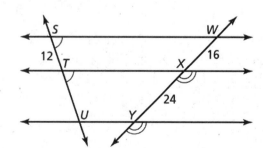

16. A reflection in a line maps point $B(3, -1)$ to point $B'(11, -1)$. What is the equation of the line of reflection?

_____

Chapter 8 Test Prep (continued)

17. Which of the following quadrilaterals does not have perpendicular diagonals?

Ⓐ rectangle

Ⓑ rhombus

Ⓒ square

Ⓓ kite

18. Which of the following statements illustrates the Transitive Property of Equality?

Ⓐ If $a = b$ and $b = c$, then $a = c$.

Ⓑ If $x = y$, then $y = x$.

Ⓒ $a = a$

Ⓓ If $AB = CD$, then $CD = AB$.

19. In a triangle, $m\angle P = 47°$ and $m\angle Q = 103°$. In another triangle, $m\angle S = (x - 7)°$ and $m\angle T = (y + 4)°$. For which of the following values of x and y are the two triangles similar?

Ⓐ $x = 37, y = 110$

Ⓑ $x = 110, y = 99$

Ⓒ $x = 54, y = 26$

Ⓓ $x = 47, y = 103$

20. The shortest side of a triangle similar to $\triangle FGH$ is 10 units long. What is the sum of the other side lengths of the triangle?

Ⓐ 12.5

Ⓑ 15

Ⓒ 20

Ⓓ 27.5

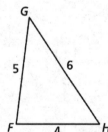

21. A carpenter cuts a piece of wood for a project. The piece of wood can be represented in the coordinate plane by a triangle with vertices $L(5, -1)$, $M(9, 7)$, and $N(1, 3)$. What type of triangle is $\triangle LMN$?

Ⓐ equilateral

Ⓑ isosceles

Ⓒ scalene

Ⓓ right

Name_____ Date_____

9.1 Extra Practice

In Exercises 1–6, find the value of *x*. Then tell whether the side lengths form a Pythagorean triple.

1.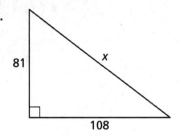

2.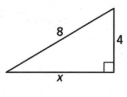

3.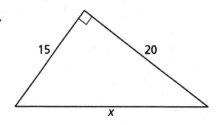

4.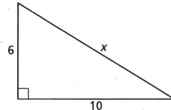

5.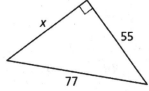

6.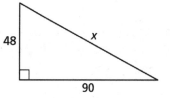

7. From school, you bike 1.2 miles due south and then 0.5 mile due east to your house. If you had biked instead on the street that runs directly from school to your house, how many fewer miles would you have biked?

8. A ski lift forms a right triangle, as shown. Use the Pythagorean Theorem to approximate the horizontal distance traveled by a person riding the ski lift. Round your answer to the nearest whole foot.

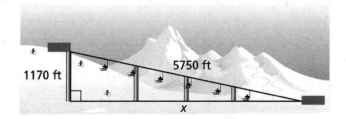

In Exercises 9–11, determine whether the segment lengths form a triangle. If so, is the triangle *acute*, *right*, or *obtuse*?

9. 90, 216, and 234

10. 1, 1, and $\sqrt{3}$

11. 4, 5, and 6

Name_____ Date_____

9.1 Review & Refresh

In Exercises 1 and 2, simplify the expression.

1. $\dfrac{18}{\sqrt{3}}$

2. $\dfrac{5}{3+\sqrt{2}}$

5. Tell whether the triangle is a right triangle.

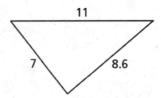

3. *ABCD* is an isosceles trapezoid, where $\overline{AB} \parallel \overline{CD}$, $\overline{AD} \cong \overline{BC}$, and $m\angle C = 105°$. Find $m\angle A$, $m\angle B$, and $m\angle D$.

6. Show that the triangles are similar and write a similarity statement.

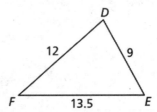 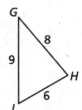

4. A company creates a triangular shelf, as shown. Determine whether $\triangle JKL \sim \triangle MKN$. Explain.

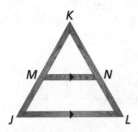

7. Graph $\triangle ABC$ with vertices $A(1, 3)$, $B(-2, 0)$, and $C(-1, 2)$ and its image after a reflection in the line $y = x$.

9.1 Self-Assessment

Use the scale to rate your understanding of the learning target and the success criteria.

1 I do not understand. **2** I can do it with help. **3** I can do it on my own. **4** I can teach someone else.

	Rating	Date
9.1 The Pythagorean Theorem		
Learning Target: Understand and apply the Pythagorean Theorem.	1 2 3 4	
I can list common Pythagorean triples.	1 2 3 4	
I can find missing side lengths of right triangles.	1 2 3 4	
I can classify a triangle as *acute*, *right*, or *obtuse* given its side lengths.	1 2 3 4	

Name_____ Date_____

9.2 Extra Practice

In Exercises 1–3, find the value of x. Write your answer in simplest form.

1.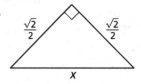
2.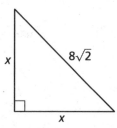
3.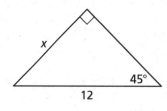

In Exercises 4–6, find the values of x and y. Write your answers in simplest form.

4.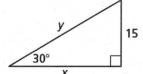
5.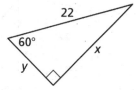
6.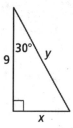

In Exercises 7 and 8, sketch the figure that is described. Find the indicated length. Write your answer in simplest form.

7. The length of a diagonal of a square is 32 inches. Find the perimeter of the square.

8. An isosceles triangle with 30° base angles has an altitude of $\sqrt{3}$ meters. Find the length of the base of the isosceles triangle.

9. Find the area of $\triangle DEF$. Write your answer in simplest form.

   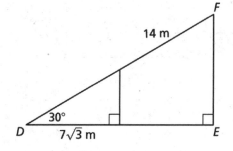

10. A 12-foot ladder is leaning up against a wall. How high does the ladder reach up the wall when x is 30°? 45°? 60°?

    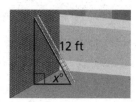

9.2 Review & Refresh

1. In the diagram, $\triangle ABC \sim \triangle XYZ$. Find the value of x.

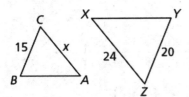

2. Determine whether segments with lengths of 4.9 meters, 7.0 meters, and 8.5 meters form a triangle. If so, is the triangle *acute*, *right*, or *obtuse*?

3. Find the values of x and y. Write your answers in simplest form.

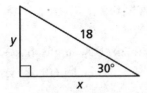

4. The endpoints of $\overline{JK}$ are $J(-4, 3)$ and $K(8, -1)$. Find the coordinates of the midpoint M.

5. Determine whether the polygons with the given vertices are congruent. Use transformations to explain your reasoning.

   $A(2, 5)$, $B(4, 6)$, $C(5, 1)$ and
   $D(-1, 4)$, $E(1, 5)$, $F(2, 0)$

6. Which tiles, if any, are similar? Explain.

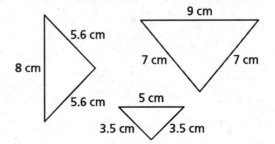

7. Three vertices of $\square WXYZ$ are $W(-3, 4)$, $Y(7, 3)$, and $Z(1, 6)$. Find the coordinates of vertex X.

8. Rewrite the definition as a biconditional statement.

   Definition A *trapezoid* is a quadrilateral with exactly one pair of parallel sides.

9.2 Self-Assessment

Use the scale to rate your understanding of the learning target and the success criteria.

| 1 I do not understand. | 2 I can do it with help. | 3 I can do it on my own. | 4 I can teach someone else. |

	Rating	Date
9.2 Special Right Triangles		
Learning Target: Understand and use special right triangles.	1 2 3 4	
I can find side lengths in 45°-45°-90° triangles.	1 2 3 4	
I can find side lengths in 30°-60°-90° triangles.	1 2 3 4	
I can use special right triangles to solve real-life problems.	1 2 3 4	

9.3 Extra Practice

In Exercises 1 and 2, identify the similar triangles.

1.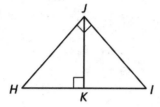

2.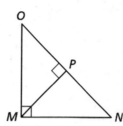

In Exercises 3–5, find the value of x.

3.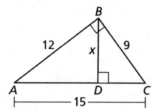

4.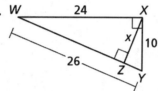

5.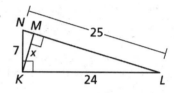

In Exercises 6–8, find the geometric mean of the two numbers.

6. 2 and 6

7. 5 and 45

8. 16 and 18

In Exercises 9–11, find the value of the variable.

9.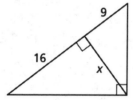

10.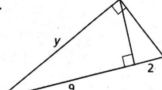

11.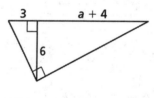

12. You are designing a kite. You know that $AB = 38.4$ centimeters, $BC = 72$ centimeters, and $AC = 81.6$ centimeters. You want to use a straight crossbar $\overline{BD}$. About how long should it be?

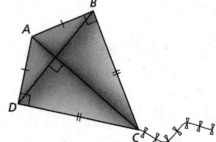

Name _____ Date _____

9.3 Review & Refresh

1. $\overline{MN}$ is a midsegment of $\triangle XYZ$. Find the value of x.

   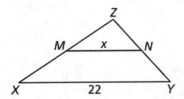

2. Find the geometric mean of 6 and 24.

3. Find the lengths of the diagonals of rectangle $ABCD$ given $AC = 6x - 5$ and $BD = 2x + 11$.

4. How tall is the tent?

   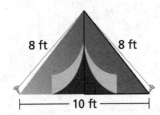

5. Tell whether the triangle is a right triangle.

   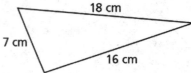

6. Find the value of y.

   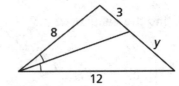

7. Determine whether the triangles are similar. If so, write a similarity statement. Explain your reasoning.

   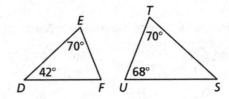

8. Graph $\triangle ABC$ with vertices $A(5, 1)$, $B(3, -2)$, and $C(2, 0)$ and its image after a translation 4 units up, followed by a reflection in the x-axis.

9.3 Self-Assessment

Use the scale to rate your understanding of the learning target and the success criteria.

| 1 I do not understand. | 2 I can do it with help. | 3 I can do it on my own. | 4 I can teach someone else. |

	Rating	Date
9.3 Similar Right Triangles		
Learning Target: Use proportional relationships in right triangles.	1 2 3 4	
I can explain the Right Triangle Similarity Theorem.	1 2 3 4	
I can find the geometric mean of two numbers.	1 2 3 4	
I can find missing dimensions in right triangles.	1 2 3 4	

Name_____ Date_____

9.4 Extra Practice

In Exercises 1–3, find the tangents of the acute angles in the right triangle. Write each answer as a fraction and as a decimal.

1.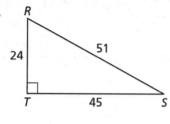

2.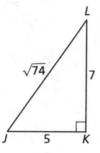

3.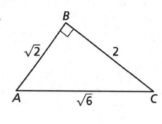

In Exercises 4–6, find the value of x.

4.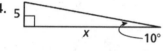

5.

6.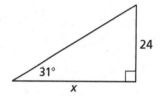

7. Describe the error in the statement of the tangent ratio. Correct the error if possible. Otherwise, write not possible.

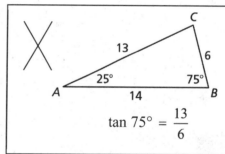

8. In $\triangle CDE$, $\angle E = 90°$ and $\tan C = \frac{4}{3}$. Find $\tan D$.

9. A boy flies a kite. The angle of elevation from his hands to the kite is 18°. The boy's hands are 4 feet above the ground. The kite is above a point that is 300 feet away from the boy. What is the height of the kite?

10. Find the perimeter of the figure.

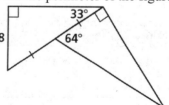

Name _____ Date _____

9.4 Review & Refresh

1. Find the value of x. Tell whether the side lengths form a Pythagorean triple.

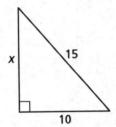

2. Find the geometric mean of 12 and 16.

3. Find the area of the polygon with vertices $(3, 1)$, $(1, 2)$, and $(3, 7)$.

4. Graph $\overline{MN}$ with endpoints $M(4, -3)$ and $N(-1, 2)$ and its image after a reflection in the y-axis, followed by a 90° rotation about the origin.

5. Find the coordinates of the circumcenter of the triangle with vertices $X(-4, 1)$, $Y(-2, 3)$, and $Z(2, -1)$.

In Exercises 6–9, find the value of x.

6.

7.

8.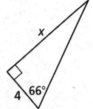

9.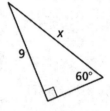

10. The vertical boards of a fence are parallel. Find $m\angle 2$.

9.4 Self-Assessment

Use the scale to rate your understanding of the learning target and the success criteria.

| 1 I do not understand. | 2 I can do it with help. | 3 I can do it on my own. | 4 I can teach someone else. |

	Rating	Date
9.4 The Tangent Ratio		
Learning Target: Understand and use the tangent ratio.	1 2 3 4	
I can explain the tangent ratio.	1 2 3 4	
I can find tangent ratios.	1 2 3 4	
I can use tangent ratios to solve real-life problems.	1 2 3 4	

9.5 Extra Practice

In Exercises 1–3, find sin F, sin G, cos F, and cos G. Write each answer as a fraction and as a decimal.

1.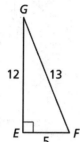
2.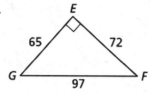
3.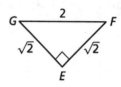

In Exercises 4–6, write the expression in terms of cosine.

4. sin 9°
5. sin 30°
6. sin 77°

In Exercises 7–9, write the expression in terms of sine.

7. cos 15°
8. cos 83°
9. cos 45°

In Exercises 10–12, find the value of each variable using sine and cosine.

10.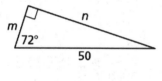
11.
12.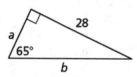

13. A camera attached to a drone is filming the damage caused by a brush fire in a closed-off area. The camera is directly above the center of the closed-off area.

   a. The person controlling the drone is standing 100 feet away from the center of the closed-off area. The person is holding the controller 3.5 feet above the ground. The angle of depression from the camera to the controller is 25°. What is the distance from the controller to the camera?

   b. The signal strength allows the drone to be 500 feet away from the controller. What is the maximum distance the person can stand from the center of the closed-off area, assuming the same angle of depression of 25°, to film the damage?

Name_____ Date_____

9.5 Review & Refresh

In Exercises 1 and 2, find the value of *x*. Tell whether the side lengths form a Pythagorean triple.

1.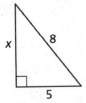
2.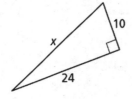

3. Write cos 71° in terms of sine.

4. Find the value of *x*.

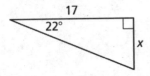

5. Find the measure of each interior angle and each exterior angle of a regular 25-gon.

6. Identify the similar right triangles. Then find the value of *x*.

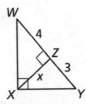

7. Find the values of *x* and *y*. Write your answers in simplest form.

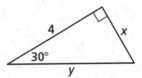

8. The polygons are congruent. Find the values of *x* and *y*.

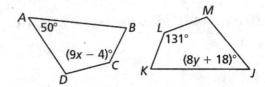

9. Draw a rectangle with a length of 12 units and a width of 2 units in a coordinate plane. Find the length of a diagonal.

10. Given the points $A(6, -1)$ and $B(0, 7)$, find the coordinates of point P along the directed line segment AB so the ratio of AP to PB is 3 to 2.

9.5 Self-Assessment

Use the scale to rate your understanding of the learning target and the success criteria.

| 1 I do not understand. | 2 I can do it with help. | 3 I can do it on my own. | 4 I can teach someone else. |

	Rating	Date
9.5 The Sine and Cosine Ratios		
Learning Target: Understand and use the sine and cosine ratios.	1 2 3 4	
I can explain the sine and cosine ratios.	1 2 3 4	
I can find sine and cosine ratios.	1 2 3 4	
I can use sine and cosine ratios to solve real-life problems.	1 2 3 4	

Name_____ Date_____

9.6 Extra Practice

In Exercises 1 and 2, determine which of the two acute angles has the given trigonometric ratio.

1. The cosine of the angle is $\frac{24}{25}$.

2. The sine of the angle is about 0.38.

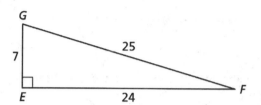
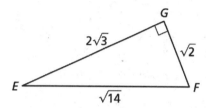

In Exercises 3–5, let ∠H be an acute angle. Use technology to approximate m∠H.

3. $\sin H = 0.2$
4. $\tan H = 1$
5. $\cos H = 0.33$

In Exercises 6–8, solve the right triangle.

6.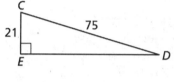
7.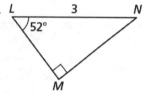
8.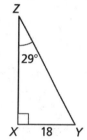

9. A boat is pulled in by a winch 12 feet above the deck of the boat. When the winch is fully extended to 25 feet, what is the angle of elevation from the deck of the boat to the winch?

10. Find the acute angle formed by Washington Boulevard and Willow Way.

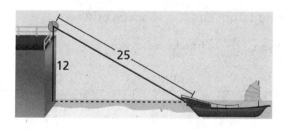

11. Find the distance across the suspension bridge.

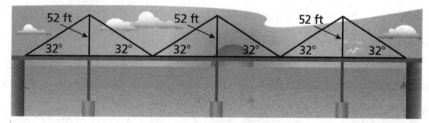

Name _____ Date _____

9.6 Review & Refresh

1. Find $\sin X$, $\cos X$, and $\tan X$. Write each answer as a fraction and as a decimal.

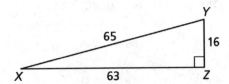

6. Find $m\angle 1$. Tell which theorem you use.

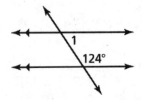

In Exercises 2 and 3, solve the proportion.

2. $\dfrac{4}{17} = \dfrac{28}{x}$

3. $\dfrac{3.3}{14.8} = \dfrac{x}{9.25}$

In Exercises 7 and 8, solve the right triangle.

7.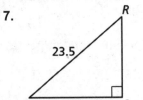

8.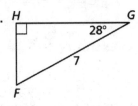

4. Identify the similar right triangles. Then find the value of y.

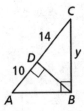

9. Determine whether the molecular model has rotational symmetry. If so, describe any rotations that map the model onto itself.

5. In the diagram, $\triangle DEF \cong \triangle XYZ$. Find the values of x and y.

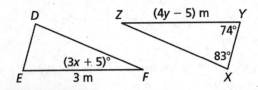

9.6 Self-Assessment

Use the scale to rate your understanding of the learning target and the success criteria.

| 1 I do not understand. | 2 I can do it with help. | 3 I can do it on my own. | 4 I can teach someone else. |

	Rating	Date
9.6 Solving Right Triangles		
Learning Target: Find unknown side lengths and angle measures of right triangles.	1 2 3 4	
I can explain inverse trigonometric ratios.	1 2 3 4	
I can use inverse trigonometric ratios to approximate angle measures.	1 2 3 4	
I can solve right triangles.	1 2 3 4	
I can solve real-life problems by solving right triangles.	1 2 3 4	

Name_____ Date_____

9.7 Extra Practice

In Exercises 1–3, use technology to find the trigonometric ratio.

1. $\sin 225°$
2. $\cos 111°$
3. $\tan 96°$

In Exercises 4 and 5, find the area of the triangle.

4.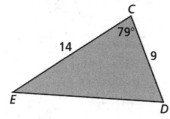

5.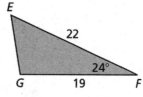

In Exercises 6–11, solve the triangle.

6.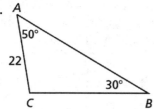

7.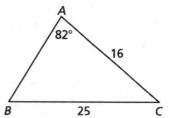

8.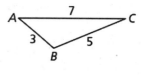

9.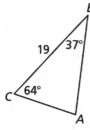

10.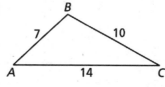

11.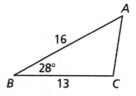

12. Determine the measures of the angles in the design of the streetlamp shown.

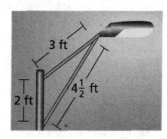

Copyright © Big Ideas Learning, LLC
All rights reserved.

Name _____ Date _____

9.7 Review & Refresh

In Exercises 1–4, find the value of x.

1.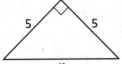

2.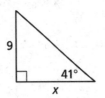

3.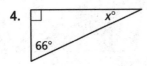

4. (triangle with right angle, 66°, and x°)

5. A triangle has one side length of 5 inches and another side length of 11 inches. Describe the possible lengths of the third side.

6. Quadrilateral $ABCD$ has vertices $A(7, 1)$, $B(5, 3)$, $C(4, 6)$, and $D(6, 4)$. Quadrilateral $EFGH$ has vertices $E(3, 3)$, $F(5, 1)$, $G(8, 0)$ and $H(6, 2)$. Are the two quadrilaterals congruent? Use transformations to explain your reasoning.

7. Solve $-4 = \sqrt[3]{11x - 20}$.

8. Find the values of x and y.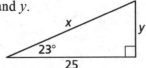

9. State which theorem you can use to show that the quadrilateral is a parallelogram.

   (parallelogram figure)

10. You draw a map of the street from your home to the post office on a coordinate plane. The post office is exactly halfway between your home and your school. You home is located at the point $(3, 7)$ and the post office is located at the point $(5, 3)$. What point represents the location of your school?

In Exercises 11 and 12, solve the triangle.

11.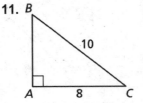

12.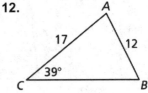

9.7 Self-Assessment

Use the scale to rate your understanding of the learning target and the success criteria.

| 1 I do not understand. | 2 I can do it with help. | 3 I can do it on my own. | 4 I can teach someone else. |

	Rating	Date
9.7 Law of Sines and Law of Cosines		
Learning Target: Find unknown side lengths and angle measures of acute and obtuse triangles.	1 2 3 4	
I can find areas of triangles using formulas that involve sine.	1 2 3 4	
I can solve triangles using the Law of Sines.	1 2 3 4	
I can solve triangles using the Law of Cosines.	1 2 3 4	

Name_____ Date_____

 Chapter Self-Assessment

Use the scale to rate your understanding of the learning target and the success criteria.

1 I do not understand. **2** I can do it with help. **3** I can do it on my own. **4** I can teach someone else.

	Rating	Date
Chapter 9 Right Triangles and Trigonometry		
Learning Target: Understand right triangles and trigonometry.	1 2 3 4	
I can use the Pythagorean Theorem to solve problems.	1 2 3 4	
I can find side lengths in special right triangles.	1 2 3 4	
I can explain how similar triangles are used with trigonometric ratios.	1 2 3 4	
I can use trigonometric ratios to solve problems.	1 2 3 4	
9.1 The Pythagorean Theorem		
Learning Target: Understand and apply the Pythagorean Theorem.	1 2 3 4	
I can list common Pythagorean triples.	1 2 3 4	
I can find missing side lengths of right triangles.	1 2 3 4	
I can classify a triangle as *acute*, *right*, or *obtuse* given its side lengths.	1 2 3 4	
9.2 Special Right Triangles		
Learning Target: Understand and use special right triangles.	1 2 3 4	
I can find side lengths in 45°-45°-90° triangles.	1 2 3 4	
I can find side lengths in 30°-60°-90° triangles.	1 2 3 4	
I can use special right triangles to solve real-life problems.	1 2 3 4	
9.3 Similar Right Triangles		
Learning Target: Use proportional relationships in right triangles.	1 2 3 4	
I can explain the Right Triangle Similarity Theorem.	1 2 3 4	
I can find the geometric mean of two numbers.	1 2 3 4	
I can find missing dimensions in right triangles.	1 2 3 4	

Chapter 9 Chapter Self-Assessment (continued)

	Rating	Date
9.4 The Tangent Ratio		
Learning Target: Understand and use the tangent ratio.	1 2 3 4	
I can explain the tangent ratio.	1 2 3 4	
I can find tangent ratios.	1 2 3 4	
I can use tangent ratios to solve real-life problems.	1 2 3 4	
9.5 The Sine and Cosine Ratios		
Learning Target: Understand and use the sine and cosine ratios.	1 2 3 4	
I can explain the sine and cosine ratios.	1 2 3 4	
I can find sine and cosine ratios.	1 2 3 4	
I can use sine and cosine ratios to solve real-life problems.	1 2 3 4	
9.6 Solving Right Triangles		
Learning Target: Find unknown side lengths and angle measures of right triangles.	1 2 3 4	
I can explain inverse trigonometric ratios.	1 2 3 4	
I can use inverse trigonometric ratios to approximate angle measures.	1 2 3 4	
I can solve right triangles.	1 2 3 4	
I can solve real-life problems by solving right triangles.	1 2 3 4	
9.7 Law of Sines and Law of Cosines		
Learning Target: Find unknown side lengths and angle measures of acute and obtuse triangles.	1 2 3 4	
I can find areas of triangles using formulas that involve sine.	1 2 3 4	
I can solve triangles using the Law of Sines.	1 2 3 4	
I can solve triangles using the Law of Cosines.	1 2 3 4	

Name_____ Date_____

Chapter 9 Test Prep

1. Which of the following is an approximation of $m\angle C$?

    Ⓐ 35.8°
    Ⓑ 43.8°
    Ⓒ 46.2°
    Ⓓ 54.2°

    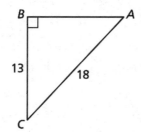

2. What is the perimeter of the parallelogram?

    Ⓐ 26 units
    Ⓑ 38 units
    Ⓒ 52 units
    Ⓓ 74 units

    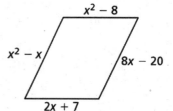

3. Which of the following are possible measures of $\angle WXZ$?

    Ⓐ 39°
    Ⓑ 51°
    Ⓒ 63°
    Ⓓ 68°
    Ⓔ 82°

4. Which proportions are true? Select all that apply.

    Ⓐ $\dfrac{AD}{AC} = \dfrac{AC}{AB}$
    Ⓑ $\dfrac{AD}{AC} = \dfrac{CD}{CB}$
    Ⓒ $\dfrac{AB}{CB} = \dfrac{BD}{DC}$
    Ⓓ $\dfrac{DC}{AC} = \dfrac{CB}{AB}$
    Ⓔ $\dfrac{AC}{AB} = \dfrac{DC}{AC}$

    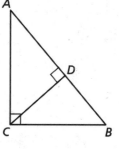

5. Which angle pair is illustrated by $\angle 5$ and $\angle 11$?

    Ⓐ corresponding angles
    Ⓑ alternate interior angles
    Ⓒ alternate exterior angles
    Ⓓ consecutive interior angles

6. The sum of the measures of the angles of a regular polygon is 2340°. What is the measure of each exterior angle?

    Ⓐ 24°
    Ⓑ 27.7°
    Ⓒ 152.3°
    Ⓓ 156°

7. Which of the following segment lengths form an acute triangle?

    Ⓐ 9, 12, 15
    Ⓑ 3, 6, 7
    Ⓒ 7, 10, 12
    Ⓓ 11, 14, 18

Chapter 9 Test Prep (continued)

8. For which of the following cases should you use the Law of Cosines?

   Ⓐ AAS case

   Ⓑ SSA case

   Ⓒ ASA case

   Ⓓ SSS case

9. $m\angle XYZ = 110°$, $m\angle XYW = (9x + 6)°$, and $m\angle WYZ = (40 - x)°$. What is $m\angle XYW$?

   Ⓐ 32°

   Ⓑ 55°

   Ⓒ 55.5°

   Ⓓ 78°

   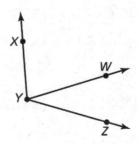

10. What is the value of x? Round to the nearest thousandth.

    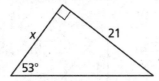

11. What is the area of the triangle? Round to the nearest hundredth.

    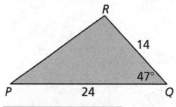

    square units

12. Complete the similarity statement.

    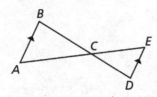

    $\triangle ABC \sim$ _____

13. Point D is the centroid of $\triangle ABC$, $DC = 8x - 6$, and $ED = 3x + 2$. What is CE?

    Ⓐ 10

    Ⓑ 15

    Ⓒ 34

    Ⓓ 51

    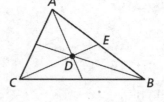

158 Geometry
Practice Workbook and Test Prep

Chapter 9 Test Prep (continued)

14. Which of the following can be concluded from the diagram?

Ⓐ $k \parallel \ell$

Ⓑ $\angle DCB$ and $\angle BCE$ form a linear pair.

Ⓒ Point C is the midpoint of $\overline{DE}$.

Ⓓ $\overrightarrow{BC} \perp \overrightarrow{DC}$

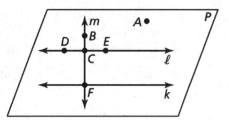

15. Which of the following transformations maps $\triangle ABC$ to $\triangle XYZ$?

Ⓐ reflection in the line $x = 4$

Ⓑ 270° rotation about the origin

Ⓒ translation 4 units right

Ⓓ reflection in the y-axis

16. What is the value of y?

Ⓐ -3

Ⓑ 4

Ⓒ 5

Ⓓ 7

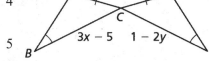

17. $\triangle FGH \sim \triangle QRS$. Which statement is false?

Ⓐ $m\angle G = m\angle R$

Ⓑ $\dfrac{FG}{GH} = \dfrac{QR}{RS}$

Ⓒ $\dfrac{FH}{QS} = \dfrac{GH}{RS}$

Ⓓ $\dfrac{GH}{RS} = \dfrac{QR}{FG}$

18. Write $\sin 76°$ in terms of cosine.

_____

19. What is $\tan B$?

Ⓐ $\dfrac{3}{2}$

Ⓑ $\dfrac{\sqrt{13}}{2}$

Ⓒ $\dfrac{3\sqrt{13}}{13}$

Ⓓ $\dfrac{\sqrt{13}}{3}$

20. What is the geometric mean of 20 and 45?

Ⓐ $\sqrt{65}$

Ⓑ 25

Ⓒ 30

Ⓓ 32.5

Chapter 9 Test Prep (continued)

21. What is the value of x?

Ⓐ $\sqrt{3}$
Ⓑ $\sqrt{6}$
Ⓒ 3
Ⓓ 6

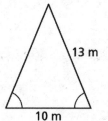

22. What is the area of the triangle?

Ⓐ 30 m²
Ⓑ 60 m²
Ⓒ 65 m²
Ⓓ 130 m²

23. You fly a kite using a 7-meter string. The angle of elevation from your hands to the kite is 54°. Your hands are 1.5 meters above the ground. How far above the ground is the kite? Round to the nearest hundredth of a meter.

_____ meters

24. Using the tick marks and arcs shown in the five triangles, which triangles are congruent to △ABC? Select all that apply.

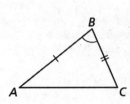

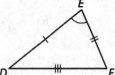

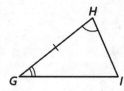

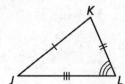

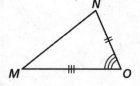

Name_____ Date_____

10.1 Extra Practice

In Exercises 1–6, use the diagram.

1. Name the circle.
2. Name two radii.
3. Name two chords.
4. Name a diameter.
5. Name a secant.
6. Name a tangent and a point of tangency.

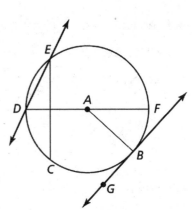

7. Tell how many common tangents the circles have and draw them. State whether the tangents are *external tangents* or *internal tangents*.

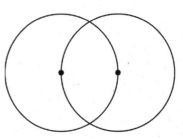

In Exercises 8–10, point B is a point of tangency. Find the radius r of ⊙C.

8.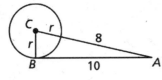
9.
10.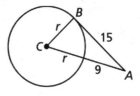

In Exercises 11–13, points B and D are points of tangency. Find the value(s) of x.

11.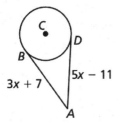
12.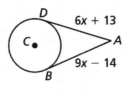
13.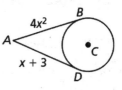

14. Two sidewalks are tangent to a circular park centered at P, as shown.

   a. What is the length of sidewalk $\overline{AB}$? Explain.

   b. What is the diameter of the park?

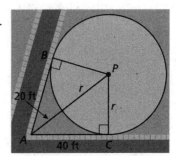

Name_____ Date_____

10.1 Review & Refresh

In Exercises 1 and 2, solve the triangle.

1.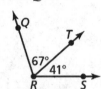
2.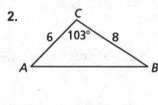

6. Point L and N are points of tangency. Find the value of x.

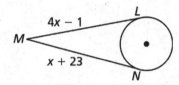

In Exercises 3 and 4, find the indicated measure.

3. $m\angle QRS$
4. YZ

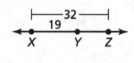

7. Find the vertical distance covered by the slide.

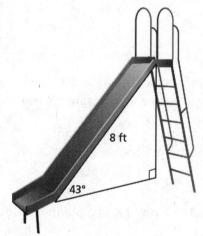

5. Tell whether the lines through the given points are *parallel*, *perpendicular*, or *neither*.

   Line 1: $(4, 7), (0, 2)$
   Line 2: $(-3, -6), (1, -1)$

8. Point P is the centroid of $\triangle ABC$. Find AP and DP when $AD = 42$.

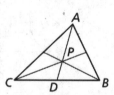

10.1 Self-Assessment

Use the scale to rate your understanding of the learning target and the success criteria.

1 I do not understand.   **2** I can do it with help.   **3** I can do it on my own.   **4** I can teach someone else.

	Rating	Date
10.1 Lines and Segments That Intersect Circles		
Learning Target: Identify lines and segments that intersect circles and use them to solve problems.	1 2 3 4	
I can identify special segments and lines that intersect circles.	1 2 3 4	
I can draw and identify common tangents.	1 2 3 4	
I can use properties of tangents to solve problems.	1 2 3 4	

162 Geometry
Practice Workbook and Test Prep

10.2 Extra Practice

In Exercises 1–6, identify the given arc as a *major arc*, *minor arc*, or *semicircle*. Then find the measure of the arc.

1. $\overset{\frown}{CD}$
2. $\overset{\frown}{AB}$
3. $\overset{\frown}{ABD}$
4. $\overset{\frown}{AEC}$
5. $\overset{\frown}{DE}$
6. $\overset{\frown}{DEB}$

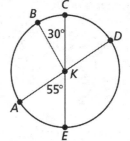

In Exercises 7–9, a recent survey asked high school students to name their favorite movie genre. The results are shown in the circle graph. Find the indicated arc measure.

7. $m\overset{\frown}{AF}$
8. $m\overset{\frown}{GDC}$
9. $m\overset{\frown}{DFC}$

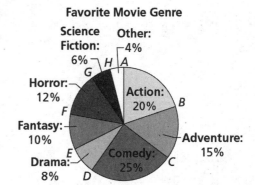

In Exercises 10 and 11, tell whether the arcs are congruent. Explain why or why not.

10. $\overset{\frown}{BC}$ and $\overset{\frown}{DE}$
11. $\overset{\frown}{WX}$ and $\overset{\frown}{YZ}$

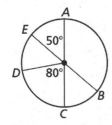
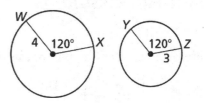

12. Find the value of x. Then find the measure of $\overset{\frown}{AB}$.

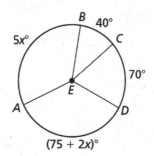

10.2 Review & Refresh

1. Points B and D are points of tangency. Find the value(s) of x.

   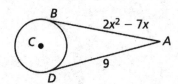

2. Name the minor arc and find its measure. Then name the major arc and find its measure.

   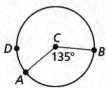

3. Find YZ.

   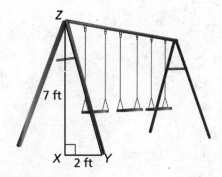

4. Find the geometric mean of 16 and 45.

In Exercises 5 and 6, find the value of x.

5.

6.

7. Graph $\triangle ABC$ with vertices $A(3, 1)$, $B(-1, 2)$, and $C(0, -2)$ and its image after a 180° rotation about the origin.

8. Find $m\angle ABC$. Explain your reasoning.

   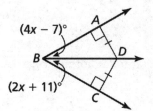

10.2 Self-Assessment

Use the scale to rate your understanding of the learning target and the success criteria.

| 1 I do not understand. | 2 I can do it with help. | 3 I can do it on my own. | 4 I can teach someone else. |

	Rating	Date
10.2 Finding Arc Measures		
Learning Target: Understand arc measures and similar circles.	1 2 3 4	
I can find arc measures.	1 2 3 4	
I can identify congruent arcs.	1 2 3 4	
I can prove that all circles are similar.	1 2 3 4	

10.3 Extra Practice

In Exercises 1–4, find the measure of the arc or chord in ⊙Q.

1. $\widehat{WX}$
2. $\overline{YZ}$
3. $\overline{WZ}$
4. $\widehat{XY}$

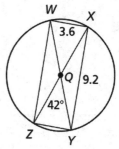

In Exercises 5 and 6, find the value of x.

5.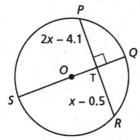

6.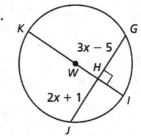

7. Three blueberry bushes are arranged, as shown. Where should you place a scarecrow so that it is the same distance from each bush?

In Exercises 8 and 9, determine whether $\overline{AB}$ is a diameter of the circle. Explain your reasoning.

8.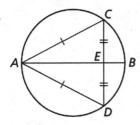

9.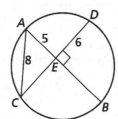

In Exercises 10 and 11, find the radius of the circle.

10.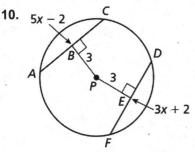

11.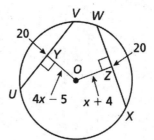

10.3 Review & Refresh

1. Point B is a point of tangency. Find the radius x of $\odot C$.

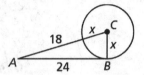

In Exercises 2 and 3, find the missing interior angle measure.

2. Quadrilateral $WXYZ$ has angle measures $m\angle W = 66°$, $m\angle X = 85°$, and $m\angle Z = 113°$. Find $m\angle Y$.

3. Pentagon $ABCDE$ has angle measures $m\angle B = 90°$, $m\angle C = 125°$, $m\angle D = 76°$, and $m\angle E = 104°$. Find $m\angle A$.

4. Tell whether $\overarc{MN}$ and $\overarc{ST}$ are congruent, similar, or neither. Explain.

 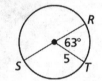

5. A surveyor makes the measurements shown to determine the length of a bridge to be built across a small lake from the West Cabins to the East Cabins. Find the length of the bridge.

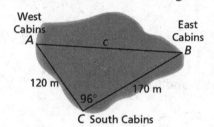

6. Find $m\overarc{EH}$ in $\odot Q$.

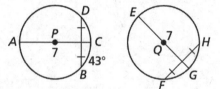

10.3 Self-Assessment

Use the scale to rate your understanding of the learning target and the success criteria.

| 1 I do not understand. | 2 I can do it with help. | 3 I can do it on my own. | 4 I can teach someone else. |

	Rating	Date
10.3 Using Chords		
Learning Target: Understand and apply theorems about chords.	1 2 3 4	
I can use chords of circles to find arc measures.	1 2 3 4	
I can use chords of circles to find lengths.	1 2 3 4	
I can describe the relationship between a diameter and a chord perpendicular to a diameter.	1 2 3 4	
I can find the center of a circle given three points on the circle.	1 2 3 4	

Name_____ Date_____

10.4 Extra Practice

In Exercises 1–6, find the indicated measure.

1. $m\angle B$

2. $m\angle E$

3. $m\widehat{HI}$

4. $m\widehat{KL}$

5. $m\widehat{MN}$

6. $m\angle U$

In Exercises 7–12, find the indicated measure.

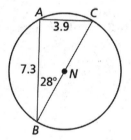

7. $m\widehat{BC}$

8. $m\angle A$

9. $m\angle C$

10. BC

11. $m\widehat{AC}$

12. $m\widehat{AB}$

13. Name two pairs of congruent angles.

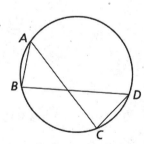

In Exercises 14 and 15, find the value of each variable.

14.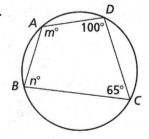

15.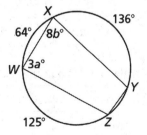

10.4 Review & Refresh

1. Describe a congruence transformation that maps △ABC to △DEF.

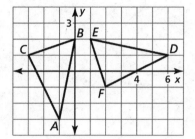

2. Find the radius of ⊙A.

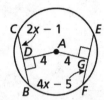

In Exercises 3–5, identify the given arc as a *major arc*, *minor arc*, or *semicircle*. Then find the measure of the arc.

3. $\widehat{YZ}$

4. $\widehat{XYZ}$

5. $\widehat{WZX}$

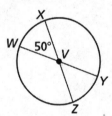

6. Tell whether $\overline{AB}$ is tangent to ⊙C. Explain your reasoning.

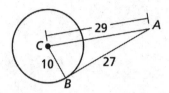

In Exercises 7 and 8, find the indicated measure.

7. m∠QRS

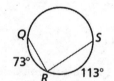

8. m∠DFE

9. You and your friend leave school heading in opposite directions. You each travel 0.7 mile, then change directions and travel 0.5 mile. You start due west and then turn 30° toward south. Your friend starts due east and then turns 15° toward north. Who is farther from the school? Explain your reasoning.

10.4 Self-Assessment

Use the scale to rate your understanding of the learning target and the success criteria.

| 1 I do not understand. | 2 I can do it with help. | 3 I can do it on my own. | 4 I can teach someone else. |

	Rating	Date
10.4 Inscribed Angles and Polygons		
Learning Target: Use properties of inscribed angles and inscribed polygons.	1 2 3 4	
I can find measures of inscribed angles and intercepted arcs.	1 2 3 4	
I can find angle measures of inscribed polygons.	1 2 3 4	
I can construct a square inscribed in a circle.	1 2 3 4	

10.5 Extra Practice

In Exercises 1–3, line *t* is tangent to the circle. Find the indicated measure.

1. $m\widehat{AB}$

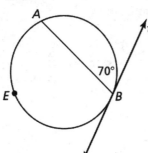

2. $m\widehat{XY}$

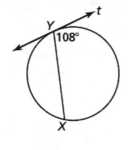

3. $m\angle 2$

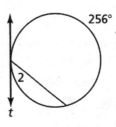

In Exercises 4–9, find the value of *x*.

4.

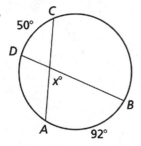

5.

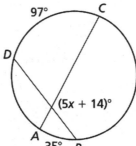

6.

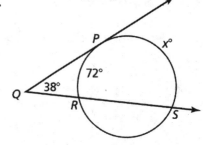

7.

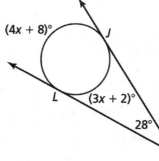

8.

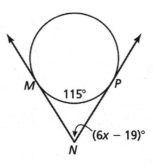

9.

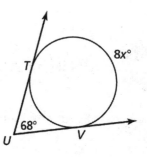

10. The diagram shows the portion of Earth visible on a clear day from the top of Mount Townsend 1.37 miles above sea level at point *B*. Find $m\widehat{CD}$.

Name_____ Date_____

10.5 Review & Refresh

1. Find the perimeter and area of the triangle with vertices $A(5, -6)$, $B(-3, -6)$, and $C(5, 7)$.

2. A diver is using sonar to track a shark that is circling his submarine. $m\angle BSC = 127°$. Find $m\overset{\frown}{AC}$. Explain your reasoning.

In Exercises 3 and 4, find the value of x.

3.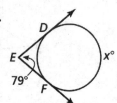

4.

5. A triangle has one side of length 5 and another side of length 24. Describe the possible lengths of the third side.

In Exercises 6 and 7, find the indicated measure.

6. $m\angle K$

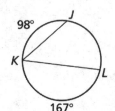

7. $m\overset{\frown}{ST}$

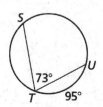

8. Graph $\triangle PQR$ with vertices $P\left(1, \frac{1}{2}\right)$, $Q\left(-\frac{3}{2}, 1\right)$, and $R\left(\frac{1}{2}, \frac{3}{2}\right)$ and its image after a dilation with a scale factor of 4.

10.5 Self-Assessment

Use the scale to rate your understanding of the learning target and the success criteria.

| 1 I do not understand. | 2 I can do it with help. | 3 I can do it on my own. | 4 I can teach someone else. |

	Rating	Date
10.5 Angle Relationships in Circles		
Learning Target: Understand angles formed by chords, secants, and tangents.	1 2 3 4	
I can identify angles and arcs determined by chords, secants, and tangents.	1 2 3 4	
I can find angle measures and arc measures involving chords, secants, and tangents.	1 2 3 4	
I can use circumscribed angles to solve problems.	1 2 3 4	

Name_____ Date_____

10.6 Extra Practice

In Exercises 1–6, find the value of x.

1.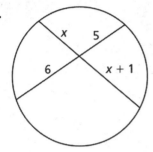

2.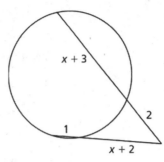

3.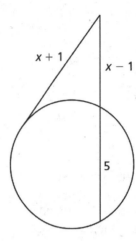

4.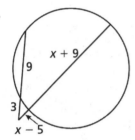

5.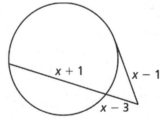

6.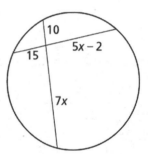

7. You go swimming in a circular pool. Your sandals are 9 feet from the ladder of the pool. The distance from the sandals to the pool is 5 feet. What is the diameter of the pool?

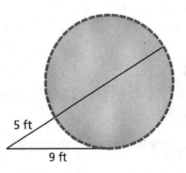

8. The Xs in the diagram show the positions of two basketball teammates relative to the free throw circle on a basketball court. The player outside the circle passes the ball to the player on the circle. What is the length of the pass?

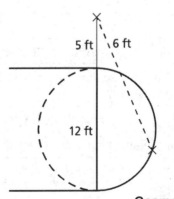

10.6 Review & Refresh

1. Find the radius of ⊙A.

   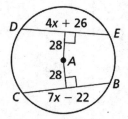

2. You measure your distance from a column and the angle of elevation from the ground to the top of the column. Find the height of the column.

In Exercises 3 and 4, find the indicated measure.

3. WY

   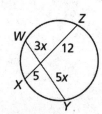

4. $m\widehat{RS}$

   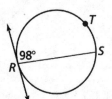

5. Find $m\widehat{UV}$.

   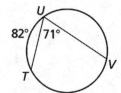

6. Show that $\triangle EGH$ and $\triangle EFG$ are similar.

   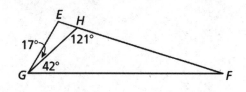

7. Find the value of x that makes $m \parallel n$. Explain your reasoning.

   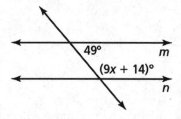

10.6 Self-Assessment

Use the scale to rate your understanding of the learning target and the success criteria.

| 1 I do not understand. | 2 I can do it with help. | 3 I can do it on my own. | 4 I can teach someone else. |

	Rating	Date
10.6 Segment Relationships in Circles		
Learning Target: Use theorems about segments of chords, secants, and tangents.	1 2 3 4	
I can find lengths of segments of chords.	1 2 3 4	
I can identify segments of secants and tangents.	1 2 3 4	
I can find lengths of segments of secants and tangents.	1 2 3 4	

10.7 Extra Practice

In Exercises 1–4, write the standard equation of the circle.

1.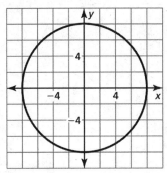

2.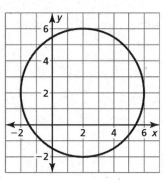

3. a circle with center $(0, 0)$ and radius $\frac{1}{3}$

4. a circle with center $(-3, -5)$ and radius 8

In Exercises 5 and 6, use the given information to write the standard equation of the circle.

5. The center is $(0, 0)$, and a point on the circle is $(4, -3)$.

6. The center is $(4, 5)$, and a point on the circle is $(0, 8)$.

In Exercises 7 and 8, find the center and radius of the circle. Then graph the circle.

7. $x^2 + y^2 + 2x + 2y = 2$

8. $x^2 + y^2 - 3x + y = \frac{5}{2}$

9. Prove or disprove that the point $(-1, 2)$ lies on the circle centered at $(-4, -1)$ with radius $3\sqrt{2}$.

10. After an earthquake, you are given seismograph readings from three locations where the coordinates are measured in miles. Find the epicenter of the earthquake.

    The epicenter is 7 miles away from $A(-3, 5)$.

    The epicenter is 4 miles away from $B(6, 0)$.

    The epicenter is 5 miles away from $C(-2, -3)$.

10.7 Review & Refresh

1. Write the standard equation of the circle with center $(-3, 5)$ that passes through the point $(-8, -7)$.

In Exercises 2 and 3, find the value of *x*.

2.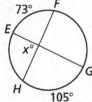
3.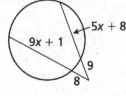

In Exercises 4–7, identify the arc as a *major arc*, *minor arc*, or *semicircle*. Then find the measure of the arc.

4. $\overarc{BDE}$

5. $\overarc{AC}$

6. $\overarc{ADC}$

7. $\overarc{CDE}$

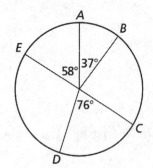

8. Find $m\angle Y$.

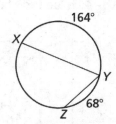

9. The lines painted for the parking spaces are parallel. Find $m\angle 2$.

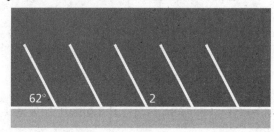

10. Find the value of *x* that makes $\triangle DEF \sim \triangle GHI$.

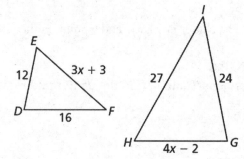

10.7 Self-Assessment

Use the scale to rate your understanding of the learning target and the success criteria.

| 1 | I do not understand. | 2 | I can do it with help. | 3 | I can do it on my own. | 4 | I can teach someone else. |

	Rating	Date
10.7 Circles in the Coordinate Plane		
Learning Target: Understand equations of circles.	1 2 3 4	
I can write equations of circles.	1 2 3 4	
I can find the center and radius of a circle.	1 2 3 4	
I can graph equations of circles.	1 2 3 4	
I can write coordinate proofs involving circles.	1 2 3 4	

10.8 Extra Practice

In Exercises 1 and 2, write an equation of the parabola.

1. focus: $(0, -8)$
   directrix: $y = 8$

2. vertex: $(0, 0)$
   focus: $(0, 1)$

In Exercises 3–5, identify the focus, directrix, and axis of symmetry of the parabola. Graph the equation.

3. $x = \frac{1}{6}y^2$

4. $x^2 = -2y$

5. $-5x + \frac{1}{3}y^2 = 0$

In Exercises 6–9, write an equation of the parabola.

6.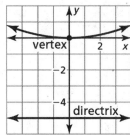

7.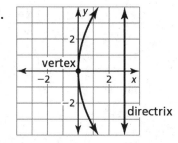

8.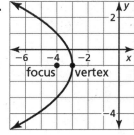

9.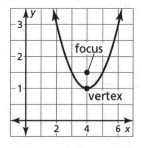

In Exercises 10 and 11, write an equation of the parabola with the given characteristics.

10. directrix: $y = 1$
    vertex: $(6, 5)$

11. focus: $\left(-\frac{15}{2}, -8\right)$
    directrix: $x = -\frac{13}{2}$

12. The cross section of a parabolic sound reflector at the Olympics has a diameter of 20 inches and a depth of 25 inches. Write an equation that represents the cross section of the reflector with its vertex at $(0, 0)$ and its focus to the left of the vertex.

13. The distance from point P to the directrix is 5 units. Write an equation of the parabola shown.

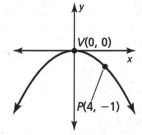

Name _____ Date _____

10.8 Review & Refresh

In Exercises 1 and 2, find the value of x.

1.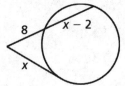
2.

3. Find the center and the radius of the circle with equation $x^2 + y^2 - 8x + 12y = 29$. Then graph the circle.

4. Write an equation of a parabola with vertex $(4, -1)$ and focus $(1, -1)$.

5. Find the measure of each angle.

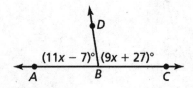

6. Triangle ABC has vertices $A(7, 3)$, $B(4, 7)$, and $C(5, 6)$. Find the circumcenter.

7. The maximum angle compass A can make is shown. Compass B is the same size as compass A but has a maximum angle of 100°. Which compass can make a larger circle? Explain.

8. Find the values of x and y.

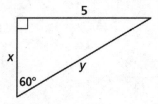

9. Graph quadrilateral $WXYZ$ with vertices $W(1, 5)$, $X(4, 2)$, $Y(3, -2)$, and $Z(0, -1)$ and its image after the composition.

   Reflection: in the line $x = 4$
   Rotation: 270° about the origin

10.8 Self-Assessment

Use the scale to rate your understanding of the learning target and the success criteria.

| 1 I do not understand. | 2 I can do it with help. | 3 I can do it on my own. | 4 I can teach someone else. |

	Rating	Date
10.8 Focus of a Parabola		
Learning Target: Graph and write equations of parabolas.	1 2 3 4	
I can explain the relationships among the focus, the directrix, and the graph of a parabola.	1 2 3 4	
I can graph parabolas.	1 2 3 4	
I can write equations of parabolas.	1 2 3 4	

Name_____ Date_____

Chapter 10 Chapter Self-Assessment

Use the scale to rate your understanding of the learning target and the success criteria.

1 I do not understand. **2** I can do it with help. **3** I can do it on my own. **4** I can teach someone else.

	Rating	Date
Chapter 10 Circles		
Learning Target: Understand and apply circle relationships.	1 2 3 4	
I can identify lines and segments that intersect circles.	1 2 3 4	
I can find angle and arc measures in circles.	1 2 3 4	
I can use circle relationships to solve problems.	1 2 3 4	
I can use circles to model and solve real-life problems.	1 2 3 4	
10.1 Lines and Segments That Intersect Circles		
Learning Target: Identify lines and segments that intersect circles and use them to solve problems.	1 2 3 4	
I can identify special segments and lines that intersect circles.	1 2 3 4	
I can draw and identify common tangents.	1 2 3 4	
I can use properties of tangents to solve problems.	1 2 3 4	
10.2 Finding Arc Measures		
Learning Target: Understand arc measures and similar circles.	1 2 3 4	
I can find arc measures.	1 2 3 4	
I can identify congruent arcs.	1 2 3 4	
I can prove that all circles are similar.	1 2 3 4	
10.3 Using Chords		
Learning Target: Understand and apply theorems about chords.	1 2 3 4	
I can use chords of circles to find arc measures.	1 2 3 4	
I can use chords of circles to find lengths.	1 2 3 4	
I can describe the relationship between a diameter and a chord perpendicular to a diameter.	1 2 3 4	
I can find the center of a circle given three points on the circle.	1 2 3 4	

Copyright © Big Ideas Learning, LLC
All rights reserved.

Geometry
Practice Workbook and Test Prep

Name _____ Date _____

Chapter 10 Chapter Self-Assessment (continued)

	Rating	Date
10.4 Inscribed Angles and Polygons		
Learning Target: Use properties of inscribed angles and inscribed polygons.	1 2 3 4	
I can find measures of inscribed angles and intercepted arcs.	1 2 3 4	
I can find angle measures of inscribed polygons.	1 2 3 4	
I can construct a square inscribed in a circle.	1 2 3 4	
10.5 Angle Relationships in Circles		
Learning Target: Understand angles formed by chords, secants, and tangents.	1 2 3 4	
I can identify angles and arcs determined by chords, secants, and tangents.	1 2 3 4	
I can find angle measures and arc measures involving chords, secants, and tangents.	1 2 3 4	
I can use circumscribed angles to solve problems.	1 2 3 4	
10.6 Segment Relationships in Circles		
Learning Target: Use theorems about segments of chords, secants, and tangents.	1 2 3 4	
I can find lengths of segments of chords.	1 2 3 4	
I can identify segments of secants and tangents.	1 2 3 4	
I can find lengths of segments of secants and tangents.	1 2 3 4	
10.7 Circles in the Coordinate Plane		
Learning Target: Understand equations of circles.	1 2 3 4	
I can write equations of circles.	1 2 3 4	
I can find the center and radius of a circle.	1 2 3 4	
I can graph equations of circles.	1 2 3 4	
I can write coordinate proofs involving circles.	1 2 3 4	
10.8 Focus of a Parabola		
Learning Target: Graph and write equations of parabolas.	1 2 3 4	
I can explain the relationships among the focus, the directrix, and the graph of a parabola.	1 2 3 4	
I can graph parabolas.	1 2 3 4	
I can write equations of parabolas.	1 2 3 4	

178 Geometry
Practice Workbook and Test Prep

Copyright © Big Ideas Learning, LLC
All rights reserved.

Name_____   Date_____

Chapter 10 Test Prep

1. Segments that are tangent to the circle form a quadrilateral. What is the perimeter of the quadrilateral?

    Ⓐ 15 m
    Ⓑ 21 m
    Ⓒ 26 m
    Ⓓ 30 m

    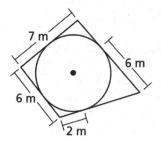

2. The measure of one interior angle of a parallelogram is 30° more than 5 times the measure of another interior angle. What is the measure of the smaller angle?

    Ⓐ 25°
    Ⓑ 30°
    Ⓒ 36°
    Ⓓ 55°

3. Write the standard equation of the circle with center $(-5, 12)$ that passes through the point $(-9, 15)$.

    _____

4. Which arc is a major arc?

    Ⓐ $\overarc{BE}$
    Ⓑ $\overarc{ADC}$
    Ⓒ $\overarc{CBA}$
    Ⓓ $\overarc{BDA}$

    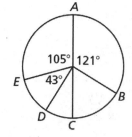

5. What is the area of the isosceles trapezoid?

    Ⓐ about 45.0 cm²
    Ⓑ about 63.2 cm²
    Ⓒ about 83.9 cm²
    Ⓓ about 94.0 cm²

6. Which congruence statements are true? Select all that apply.

    Ⓐ $\overline{KN} \cong \overline{JN}$
    Ⓑ $\angle JLM \cong \angle MKJ$
    Ⓒ $\overline{MN} \cong \overline{LN}$
    Ⓓ $\overline{KM} \cong \overline{LJ}$
    Ⓔ $\triangle LMN \cong \triangle LKJ$
    Ⓕ $\triangle KLM \cong \triangle LKJ$

    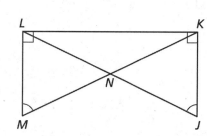

Chapter 10 Test Prep (continued)

7. What is the radius of ⊙E?

   Ⓐ 21
   Ⓑ 29
   Ⓒ 42
   Ⓓ 46.5

   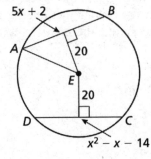

8. What is the perimeter of △ABC?

   Ⓐ about 24.8 ft
   Ⓑ about 28.4 ft
   Ⓒ about 34.0 ft
   Ⓓ about 37.5 ft

   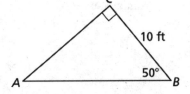

9. $\overline{BE}$ is a midsegment of △ACD. What is the perimeter of trapezoid BCDE?

   Ⓐ 14 m
   Ⓑ 20 m
   Ⓒ 28 m
   Ⓓ 31 m

10. In trapezoid ABCD, $\overline{AB} \parallel \overline{CD}$, CD = 4 • AB, and $\overline{MN}$ is the midsegment of ABCD. What is the ratio of CD to MN?

    Ⓐ 2 : 5
    Ⓑ 5 : 8
    Ⓒ 8 : 5
    Ⓓ 5 : 2

11. Which statements guarantee that △QRS ~ △XYZ? Select all that apply.

    Ⓐ △QRS and △XYZ are equilateral.
    Ⓑ m∠Q = 39°, m∠R = 56°, m∠X = 39°, and m∠Z = 95°.
    Ⓒ △XYZ is a right isosceles triangle and m∠Q + m∠R = 90°.
    Ⓓ m∠R = 107°, m∠S = 25°, m∠X = 48°, and m∠Y = 107°.
    Ⓔ m∠Q + m∠S = 138° and m∠Y + m∠Z = 116°.

12. What is the value of x?

    Ⓐ 42
    Ⓑ 48
    Ⓒ 84
    Ⓓ 96

    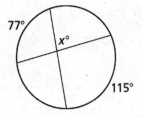

13. What is m∠QUT?

    Ⓐ 42°
    Ⓑ 54.3°
    Ⓒ 85.2°
    Ⓓ 138°

    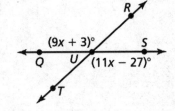

Chapter 10 Test Prep (continued)

14. What is the difference of the geometric mean and the arithmetic mean of 18 and 128?

15. What is the value of x rounded to the nearest thousandth?

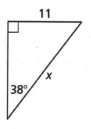

16. What is the value of y?

Ⓐ 21
Ⓑ 23
Ⓒ 42
Ⓓ 46

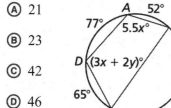

17. What is $m\widehat{QS}$?

Ⓐ 25
Ⓑ 32
Ⓒ 43
Ⓓ 52

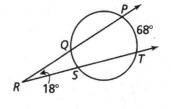

18. Point B is a point of tangency. What is the radius r of $\odot C$?

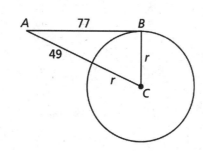

Chapter 10 Test Prep (continued)

19. What is AB?

Ⓐ 2
Ⓑ 5
Ⓒ 8
Ⓓ 16

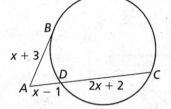

20. What is the radius of the circle?

Ⓐ 10.4
Ⓑ 14.3
Ⓒ 20.8
Ⓓ 28.6

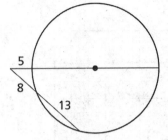

21. You graph the circle $(x+3)^2 + (y-2)^2 = 25$ and the line $x = -8$ in a coordinate plane. Which statement is true?

Ⓐ The line is a tangent of the circle.
Ⓑ The line is a secant of the circle.
Ⓒ The line is a secant that contains the diameter of the circle.
Ⓓ The line does not pass through the circle.

22. Write an equation of the parabola with vertex $(6, 1)$ and directrix $x = 3$.

_____

23. What is $m\angle R$?

Ⓐ 24°
Ⓑ 28°
Ⓒ 48°
Ⓓ 96°

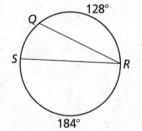

24. What is $m\angle A$?

Ⓐ 12.25°
Ⓑ 23.5°
Ⓒ 38.75°
Ⓓ 51.25°

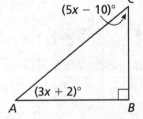

25. $\triangle XYZ$ has vertices $X(-2, 1)$, $Y(3, -1)$, and $Z(1, 4)$. What are the vertices of its image after a dilation with a scale factor of 4?

_____

Name_____ Date_____

11.1 Extra Practice

In Exercises 1–7, find the indicated measure.

1. diameter of a circle with a circumference of 10 inches

2. circumference of a circle with a radius of 3 centimeters

3. radius of a circle with a circumference of 8 feet

4. circumference of a circle with a diameter of 2.4 meters

5. arc length of $\overset{\frown}{AC}$

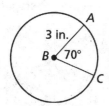

6. $m\overset{\frown}{CD}$

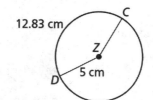

7. radius of $\odot E$

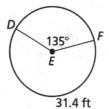

In Exercises 8 and 9, find the perimeter of the figure.

8.

9.

In Exercises 10 and 11, convert the angle measure.

10. Convert 60° to radians.

11. Convert $\dfrac{5\pi}{6}$ radians to degrees.

12. A carousel has a diameter of 50 feet.

   a. To the nearest foot, how far does a child seated near the outer edge travel when the carousel makes 8 revolutions?

   b. The carousel makes 14 revolutions. To the nearest foot, how much farther does a child seated near the outer edge travel than a child seated 15 feet from the center of the carousel?

Name _____ Date _____

11.1 Review & Refresh

In Exercises 1 and 2, find the area of the polygon with the given vertices.

1. $A(-4, 7), B(8, 7), C(-4, 2)$

2. $W(3, 8), X(3, -1), Y(-5, -1), Z(-5, 8)$

3. Find the length of the midsegment of the trapezoid.

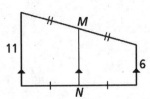

In Exercises 4 and 5, find the center and radius of the circle. Then graph the circle.

4. $x^2 + y^2 = 9$ 5. $x^2 + y^2 - 8x = 20$

In Exercises 6 and 7, find the value of x.

6. 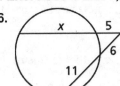   7.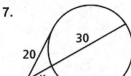

8. The Tyne Bridge is a parabolic truss arch bridge located in Newcastle, England. The outermost arch can be represented by a parabola with focus $(87, 22)$ and vertex $(87, 56)$. Write an equation of the parabola.

11.1 Self-Assessment

Use the scale to rate your understanding of the learning target and the success criteria.

| 1 I do not understand. | 2 I can do it with help. | 3 I can do it on my own. | 4 I can teach someone else. |

	Rating	Date
11.1 Circumference and Arc Length		
Learning Target: Understand circumference, arc length, and radian measure.	1 2 3 4	
I can use the formula for the circumference of a circle to find measures.	1 2 3 4	
I can find arc lengths and use arc lengths to find measures.	1 2 3 4	
I can solve real-life problems involving circumference.	1 2 3 4	
I can explain radian measure and convert between degree and radian measure.	1 2 3 4	

Name_____ Date_____

11.2 Extra Practice

In Exercises 1–4, find the indicated measure.

1. area of a circle with a diameter of 1.8 inches

2. diameter of a circle with an area of 10 square feet

3. radius of a circle with an area of 65 square centimeters

4. area of a circle with a radius of 6.1 yards

In Exercises 5–7, find the areas of the sectors formed by ∠PQR.

5.

6.

7.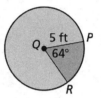

In Exercises 8 and 9, the area of the shaded sector is shown. Find the indicated measure.

8. area of ⊙Y

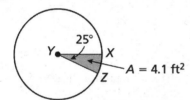

9. radius of ⊙Y

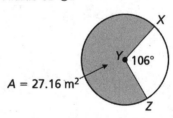

In Exercises 10–12, find the area of the shaded region.

10.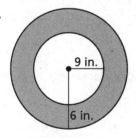

11.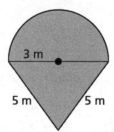

12.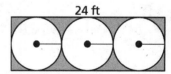

13. The diagram shows the area covered by a security camera outside a building. A new security camera is installed in the same position that doubles the radius of the coverage area. How does doubling the radius affect the coverage area? Explain.

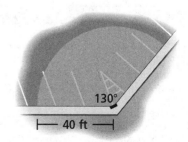

Name _____ Date _____

11.2 Review & Refresh

1. Find the area of the shaded region.

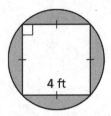

In Exercises 2 and 3, find the indicated measure.

2. $m\widehat{AB}$

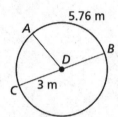

3. $m\angle MQN$

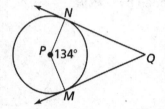

4. Graph $x = \frac{1}{16}(y-3)^2 + 2$. Identify the focus, directrix, and axis of symmetry.

5. Find the distance from the point $(-1, -2)$ to the line $y = \frac{1}{2}x + 6$.

6. Point D is the centroid of $\triangle ABC$. Find DE and BE when $BD = 18$.

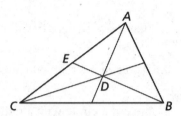

7. A *pediment* is the upper triangular part of a façade of classical architecture. Prove that the lengths of both sides of the pediment are the same.

8. Find the lengths of the diagonals of rectangle $WXYZ$ when $WY = 6x - 1$ and $XZ = 8x - 7$.

9. The endpoints of $\overline{MN}$ are $M(-6, 1)$ and $N(-2, 9)$. Write an equation of the perpendicular bisector of $\overline{MN}$.

11.2 Self-Assessment

Use the scale to rate your understanding of the learning target and the success criteria.

| 1 I do not understand. | 2 I can do it with help. | 3 I can do it on my own. | 4 I can teach someone else. |

	Rating	Date
11.2 Areas of Circles and Sectors		
Learning Target: Find areas of circles and areas of sectors of circles.	1 2 3 4	
I can use the formula for area of a circle to find measures.	1 2 3 4	
I can find areas of sectors of circles.	1 2 3 4	
I can solve problems involving areas of sectors.	1 2 3 4	

Name_____ Date_____

11.3 Extra Practice

In Exercises 1 and 2, find the area of the kite or rhombus.

1.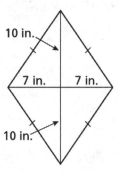

2.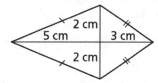

3. Find the measure of a central angle of a regular octagon.

4. Each central angle of a regular polygon is 40°. How many sides does the polygon have?

In Exercises 5–9, find the area of the regular polygon.

5.

6.

7.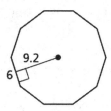

8. a pentagon with a radius of 4 units

9. a hexagon with an apothem of 10 units

In Exercises 10–12, find the area of the shaded region.

10.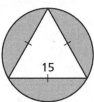

11.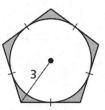

12.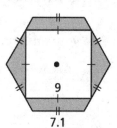

13. Use the figure of the gazebo.

    a. An arm rail is built along the perimeter of the gazebo. What is the length of the arm rail?

    b. A container of wood sealer covers 200 square feet. How many containers of sealer do you need to cover the entire floor of the gazebo? Explain your reasoning.

Name _____ Date _____

11.3 Review & Refresh

In Exercises 1 and 2, find the indicated measure.

1. area of ⊙F

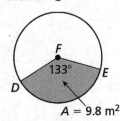

2. arc length of $\widehat{JK}$

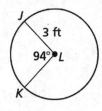

7. Find the area of the regular hexagon.

3. Write an equation of a parabola with focus $F(2, 7)$ and directrix $y = 1$.

8. An eraser is shown.

   a. State which theorem you can use to show that the side of the eraser is in the shape of a parallelogram.

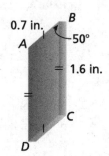

4. △ABC has vertices $A(-1, 3)$, $B(-1, -1)$, and $C(0, 1)$. △DEF has vertices $D(0, 5)$, $E(0, -3)$, and $F(2, 1)$. Are the triangles similar? Use transformations to explain your reasoning.

   b. Find $m\angle A$, $m\angle C$, and $m\angle D$.

9. Find the measure of the exterior angle.

In Exercises 5 and 6, find the value of x.

5.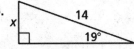

6.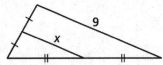

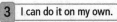

11.3 Self-Assessment

Use the scale to rate your understanding of the learning target and the success criteria.

| 1 I do not understand. | 2 I can do it with help. | 3 I can do it on my own. | 4 I can teach someone else. |

	Rating	Date
11.3 Areas of Polygons		
Learning Target: Find angle measures and areas of regular polygons.	1 2 3 4	
I can find areas of rhombuses and kites.	1 2 3 4	
I can find angle measures in regular polygons.	1 2 3 4	
I can find areas of regular polygons.	1 2 3 4	
I can explain how the area of a triangle is related to the area formulas for rhombuses, kites, and regular polygons.	1 2 3 4	

11.4 Extra Practice

In Exercises 1–3, find the indicated measure.

1. The state of Wyoming has a population of about 567,025 people. Find the population density in people per square mile.

2. About 70,000 people live in a circular region with a 30-mile radius. Find the population density in people per square mile.

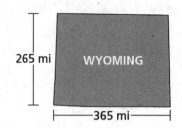

3. A circular region with a 2.8-mile radius has a population density of 1405 people per square mile. Find the number of people who live in the region.

In Exercises 4 and 5, find the radius of the region.

4. About 1.81 million people live in a circular region with a population density of about 9232 people per square kilometer.

5. About 109,000 people live in a circular region with a population density of about 743 people per square mile.

6. You are designing a rectangular dog park with an area of 21,904 square feet. The city wants you to minimize the amount of fencing used along the perimeter of the park. Each foot of fencing costs $31 to install. The gate among the fencing is 3 feet wide and costs $120 to install. How much will it cost to install the dog park fence including the gate?

7. The 8-inch segment of magnetic tape stores 12 gigabits of data. Find the areal density in bits per square inch.

8. Sun City, Arizona is a retirement community and consists of circular developments.

   a. In 2010, the population of Sun City was 37,499 people. The population density was 2610.4 people per square mile. What is the area of Sun City?

   b. Part of the community was developed as a circle with a radius of about 0.37 mile. About how many people would you expect to live in this region?

11.4 Review & Refresh

In Exercises 1 and 2, find the indicated measure.

1. $m\overset{\frown}{WXY}$

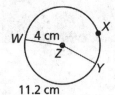

2. area of each sector

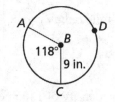

3. In the diagram, *ABCDEFG* is a regular heptagon inscribed in $\odot H$. The radius of the circle is 5 units. Find the area of the heptagon.

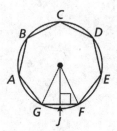

4. About 545,000 people live in a circular region with a population density of about 1175 people per square mile. Find the radius of the region.

5. A school swimming pool is being remodeled. The pool will be similar to an Olympic-size pool, which has a length of 50 meters and a width of 25 meters. The school plans to make the length of the new pool 40 meters. Find the perimeters of an Olympic-size pool and the new pool.

6. How does *JL* compare to *SQ*? Explain your reasoning.

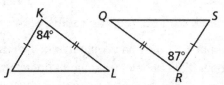

7. Find the value of each variable using sine and cosine.

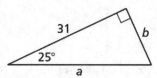

8. Find the sum of the measures of the interior angles of a 23-gon.

9. Find the geometric mean of 32 and 50.

11.4 Self-Assessment

Use the scale to rate your understanding of the learning target and the success criteria.

1 I do not understand.   2 I can do it with help.   3 I can do it on my own.   4 I can teach someone else.

	Rating	Date
11.4 Modeling with Area		
Learning Target: Understand the concept of population density and modeling with area.	1 2 3 4	
I can explain what population density means.	1 2 3 4	
I can find and use population densities.	1 2 3 4	
I can use area formulas to solve problems.	1 2 3 4	

Name_____ Date_____

Chapter 11 Chapter Self-Assessment

Use the scale to rate your understanding of the learning target and the success criteria.

1 I do not understand. **2** I can do it with help. **3** I can do it on my own. **4** I can teach someone else.

	Rating	Date
Chapter 11 Circumference and Area		
Learning Target: Understand circumference and area.	1 2 3 4	
I can find circumferences of circles and arc lengths of sectors.	1 2 3 4	
I can find areas of circles and sectors.	1 2 3 4	
I can find areas of polygons.	1 2 3 4	
I can solve real-life problems involving area.	1 2 3 4	
11.1 Circumference and Arc Length		
Learning Target: Understand circumference, arc length, and radian measure.	1 2 3 4	
I can use the formula for the circumference of a circle to find measures.	1 2 3 4	
I can find arc lengths and use arc lengths to find measures.	1 2 3 4	
I can solve real-life problems involving circumference.	1 2 3 4	
I can explain radian measure and convert between degree and radian measure.	1 2 3 4	
11.2 Areas of Circles and Sectors		
Learning Target: Find areas of circles and areas of sectors of circles.	1 2 3 4	
I can use the formula for area of a circle to find measures.	1 2 3 4	
I can find areas of sectors of circles.	1 2 3 4	
I can solve problems involving areas of sectors.	1 2 3 4	
11.3 Areas of Polygons		
Learning Target: Find angle measures and areas of regular polygons.	1 2 3 4	
I can find areas of rhombuses and kites.	1 2 3 4	
I can find angle measures in regular polygons.	1 2 3 4	
I can find areas of regular polygons.	1 2 3 4	
I can explain how the area of a triangle is related to the area formulas for rhombuses, kites, and regular polygons.	1 2 3 4	

Name _____ Date _____

 Chapter Self-Assessment (continued)

	Rating	Date
11.4 Modeling with Area		
Learning Target: Understand the concept of population density and modeling with area.	1 2 3 4	
I can explain what population density means.	1 2 3 4	
I can find and use population densities.	1 2 3 4	
I can use area formulas to solve problems.	1 2 3 4	

Name_____ Date_____

Chapter 11 Test Prep

1. About 1.28 million people live in a circular region with an 18-mile radius. What is the population density in people per square mile?

   Ⓐ 1258
   Ⓑ 3951
   Ⓒ 5361
   Ⓓ 10,722

2. A 60° arc in ⊙A and a 24° arc in ⊙B have the same length. What is the ratio of the radius of ⊙A to the radius of ⊙B?

   Ⓐ 1 to 15
   Ⓑ 1 to 6
   Ⓒ 2 to 5
   Ⓓ 5 to 2

3. Point M lies on $\overline{AB}$, $AB = 24$, and $BM = 13$. What is AM?

   Ⓐ 11
   Ⓑ 12
   Ⓒ 24
   Ⓓ 37

4. What is the area of a regular nonagon with an apothem of 4 units?

   Ⓐ about 23.1 square units
   Ⓑ about 26.2 square units
   Ⓒ about 46.3 square units
   Ⓓ about 52.4 square units

5. What is the value of x that makes $m \parallel n$?

   Ⓐ $1\frac{1}{3}$
   Ⓑ 8
   Ⓒ 20
   Ⓓ 40

   (18x − 4)° → m
   (3x + 16)° → n

6. What is $m\overset{\frown}{AC}$?

   Ⓐ 113°
   Ⓑ 137°
   Ⓒ 156°
   Ⓓ 223°

   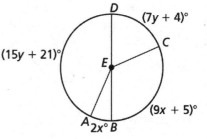

7. ⊙$X \cong$ ⊙Y. Which of the following statements are true? Select all that apply.

   Ⓐ $GH = 10$
   Ⓑ $AE = 5$
   Ⓒ $m\overset{\frown}{AD} = 106°$
   Ⓓ $m\overset{\frown}{FH} = 148°$
   Ⓔ $m\angle FHG = 74°$
   Ⓕ $GY = 6.3$

   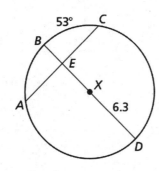 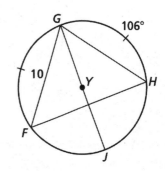

Chapter 11 Test Prep (continued)

8. Which graph represents the equation $x^2 + y^2 - 6x + 2y = 6$?

Ⓐ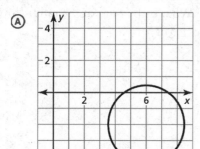

Ⓑ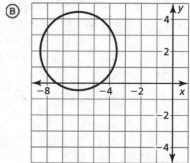

Ⓒ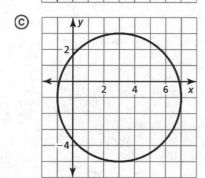

Ⓓ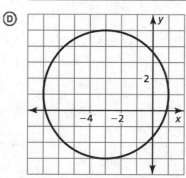

9. What is the value of x?

Ⓐ 110
Ⓑ 120
Ⓒ 140
Ⓓ 150

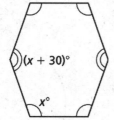

10. What is the area of the rhombus?

Ⓐ 65
Ⓑ 120
Ⓒ 130
Ⓓ 240

11. Which statements contradict each other?

I. $\triangle ABC$ is a right triangle.

II. $m\angle A = 41°$

III. $m\angle B = 103°$

Ⓐ I and II

Ⓑ I and III

Ⓒ II and III

Ⓓ None of the statements are contradictory.

12. Write an equation of the parabola.

_____

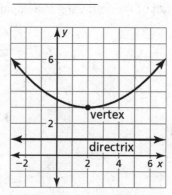

194 Geometry
Practice Workbook and Test Prep

Chapter 11 Test Prep (continued)

13. What is the perimeter of the triangle?

Ⓐ $8\sqrt{3}$ cm

Ⓑ $12 + 4\sqrt{3}$ cm

Ⓒ $8 + 8\sqrt{2}$ cm

Ⓓ $24 + 8\sqrt{3}$ cm

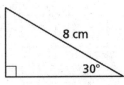

14. What is the radius of $\odot C$?

Ⓐ 10 in.

Ⓑ 17.7 in.

Ⓒ 36 in.

Ⓓ 63.8 in.

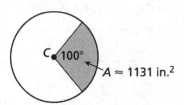

15. What is the area of trapezoid *BDEG*?

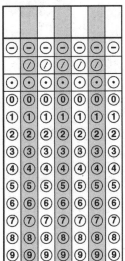 square units

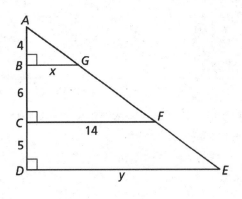

16. What is the circumference of a circle with a diameter of 24 feet?

Ⓐ 12π ft

Ⓑ 24π ft

Ⓒ 144π ft

Ⓓ 576π ft

17. What is the diameter of a circle with an area of 676π square meters?

Ⓐ 13 m

Ⓑ 26 m

Ⓒ 52 m

Ⓓ 338 m

18. You have 22.4 yards of fencing to enclose a rectangular garden. What is the maximum area of the garden?

_____

Chapter 11 Test Prep (continued)

19. A regular octagon has a perimeter of 41.6 meters. What is the radius rounded to the nearest hundredth?

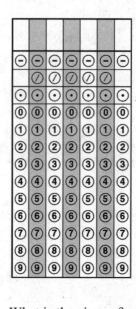
m

20. What is the area of the shaded region rounded to the nearest hundredth?

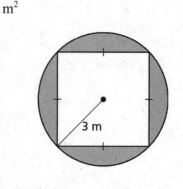
m²

21. What is the measure of the angle between the x-axis and the line $y = \frac{8}{9}x$?

Ⓐ about 27.3°
Ⓑ about 41.6°
Ⓒ about 48.4°
Ⓓ about 62.7°

22. What is the circumference of the circle $(x-2)^2 + (y+5)^2 = 81$?

Ⓐ 9π
Ⓑ 18π
Ⓒ 81π
Ⓓ 162π

23. $\triangle MNP \cong \triangle STU$. Which of the following statements are true? Select all that apply.

Ⓐ $\angle M \cong \angle S$
Ⓑ $\overline{NP} \cong \overline{ST}$
Ⓒ $\triangle PMN \cong \triangle UST$
Ⓓ $\triangle NMP \cong \triangle TUS$
Ⓔ $\overline{MP} \cong \overline{SU}$
Ⓕ $\angle P \cong \angle T$

24. Which of the following are vertices of the image of $\triangle XYZ$ after a reflection in the y-axis and a dilation with a scale factor of 2? Select all that apply.

Ⓐ $(-8, 4)$
Ⓑ $(2, -4)$
Ⓒ $(-10, 8)$
Ⓓ $(8, 4)$
Ⓔ $(2, 4)$
Ⓕ $(10, 8)$

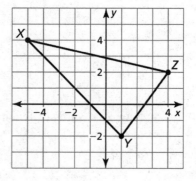

196 Geometry
Practice Workbook and Test Prep

Name_____ Date_____

12.1 Extra Practice

In Exercises 1–3, tell whether the solid is a polyhedron. If it is, name the polyhedron.

1.
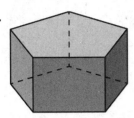

2.

3.
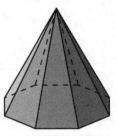

In Exercises 4–7, describe the shape formed by the intersection of the plane and the solid.

4.
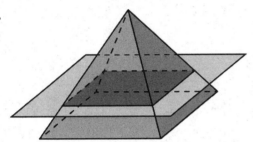

5.
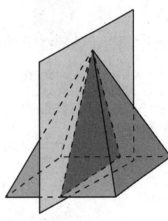

6.
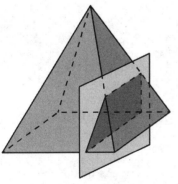

7.
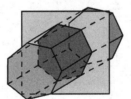

8. Consider the rectangular prism shown. The length of the prism is 5 inches, the width is 3 inches, and the height is 2 inches.

   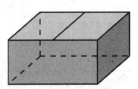

   a. Draw the cross section formed by the plane perpendicular to the base that contains the line segment drawn on the solid. What is the shape of the cross section?

   b. What is the perimeter of the cross section?

   c. What is the area of the cross section?

Name _____ Date _____

12.1 Review & Refresh

1. Explain how to prove that $\overline{QR} \cong \overline{TS}$.

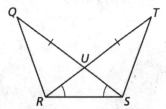

2. Tailors want to know the density of fabric when deciding what material to use when making clothing. The piece shown weighs 3 ounces. Find the density of the fabric in ounces per square yard.

9 in.
24 in.

In Exercises 3 and 4, draw the cross section formed by the described plane that contains the line segment drawn on the solid. What is the shape of the cross section?

3. plane is perpendicular to base

4. plane is parallel to base

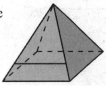

5. Tell whether $\overline{AB}$ is tangent to $\odot C$. Explain your reasoning.

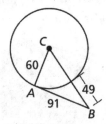

6. Three vertices of $\square WXYZ$ are $W(-5, 1)$, $Y(4, -2)$, and $Z(1, 2)$. Find the coordinates of the remaining vertex.

7. Solve the right triangle.

![triangle DEF with DF=26, angle D=19°, right angle at E]

8. Verify that the segment lengths 29, 34, and 51 form a triangle. Is the triangle *acute*, *right*, or *obtuse*?

In Exercises 9 and 10, find the area of the quadrilateral.

9.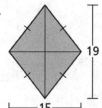
19
15

10.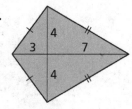
3, 4, 4, 7

12.1 Self-Assessment

Use the scale to rate your understanding of the learning target and the success criteria.

| 1 I do not understand. | 2 I can do it with help. | 3 I can do it on my own. | 4 I can teach someone else. |

	Rating	Date
12.1 Cross Sections of Solids		
Learning Target: Describe and draw cross sections.	1 2 3 4	
I can describe attributes of solids.	1 2 3 4	
I can describe and draw cross sections.	1 2 3 4	
I can solve real-life problems involving cross sections.	1 2 3 4	

198 Geometry
Practice Workbook and Test Prep

Name_____   Date_____

12.2 Extra Practice

In Exercises 1–3, find the volume of the prism.

1. Area of base 10 in.²   4 in.

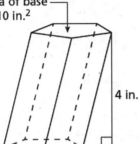

2. 4 ft, 5 ft, 14 ft, 10 ft

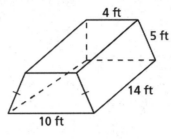

3. 1.5 cm, 5 cm, 3 cm

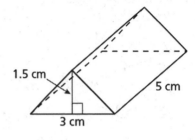

In Exercises 4–6, find the volume of the cylinder.

4. 2.5 in., 7 in.

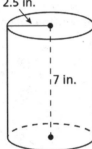

5. 2 cm, 11 cm

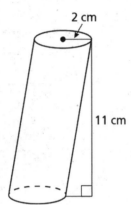

6. 7 cm, 23 cm, 70°

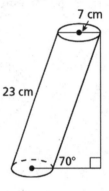

In Exercises 7–9, find the missing dimension of the prism or cylinder.

7. Volume = 75.36 in.³, 6 in., p

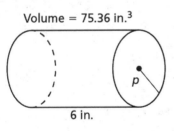

8. Volume = 40 yd³, 5 yd, 8 yd, x

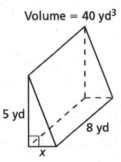

9. Volume = 661.5 cm³, 3.5 cm, 18 cm, y

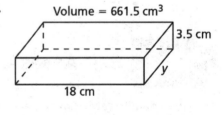

10. Find the volume of the composite solid.

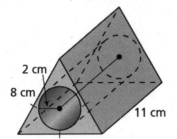

2 cm, 8 cm, 11 cm

11. The solids are similar. Find the surface area and volume of cylinder B.

Cylinder A   Cylinder B
10 m   15 m

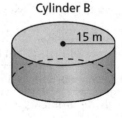

$S = 336\pi$ m²
$V = 680\pi$ m³

Copyright © Big Ideas Learning, LLC   Geometry **199**
All rights reserved.   Practice Workbook and Test Prep

Name _____ Date _____

12.2 Review & Refresh

1. In the diagram, $ABCD \cong EFGH$. Find the values of x and y.

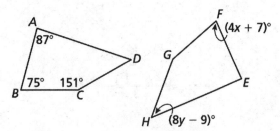

2. Find the value of x.

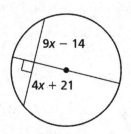

In Exercises 2 and 3, tell whether the solid is a polyhedron. If it is, name the polyhedron.

2.

3.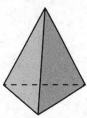

6. An airplane travels 400 miles east, then turns 50° toward north and travels another 550 miles. How far is the airplane from its starting location?

In Exercises 7 and 8, find the volume of the prism or cylinder.

7.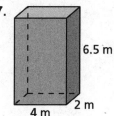

8.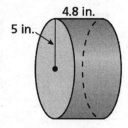

4. In $\triangle PQR$, $m\angle Q = 27°$ and $m\angle R = 79°$. In $\triangle STU$, $m\angle S = 74°$ and $m\angle T = 27°$. Are the triangles similar? Explain.

9. Find the area of the regular polygon.

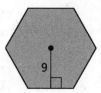

12.2 Self-Assessment

Use the scale to rate your understanding of the learning target and the success criteria.

| 1 I do not understand. | 2 I can do it with help. | 3 I can do it on my own. | 4 I can teach someone else. |

	Rating	Date
12.2 Volumes of Prisms and Cylinders		
Learning Target: Find and use volumes of prisms and cylinders.	1　2　3　4	
I can find volumes of prisms and cylinders.	1　2　3　4	
I can find surface areas and volumes of similar solids.	1　2　3　4	
I can solve real-life problems involving volumes of prisms and cylinders.	1　2　3　4	

200 Geometry
Practice Workbook and Test Prep

Name_____ Date_____

12.3 Extra Practice

In Exercises 1–3, find the volume of the pyramid.

1.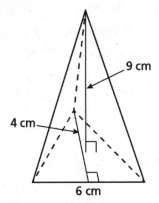

2.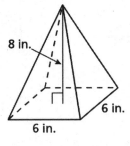

3.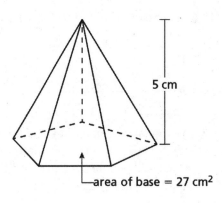

4. A pyramid with an isosceles triangular base has a volume of 18 cubic centimeters and a height of 9 centimeters. The triangular base has a height of 3 centimeters. Find the perimeter of the triangular base.

In Exercises 5 and 6, find the height of the pyramid.

5. Volume = 48 yd³

   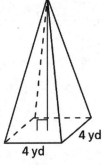

6. Volume = 8 ft³

   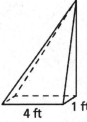

7. The pyramids are similar. Find the volume of pyramid B.

   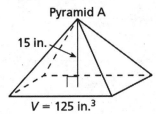
   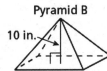

8. Find the volume of the composite solid.

9. The Pyramid Arena in Memphis, Tennessee is about 98 meters tall and has a square base with a side length of about 180 meters. A building has the shape of a rectangular prism and has a square base with the same dimensions as the Pyramid Arena. What is the height of the building when it has the same volume as the Pyramid Arena?

12.3 Review & Refresh

In Exercises 1 and 2, find the value of x.

1.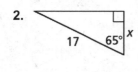

2.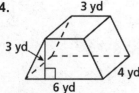

3. A circular region has a population of about 150,000 people and a population density of about 477 people per square mile. Find the radius of the region.

In Exercises 4 and 5, find the volume of the prism.

4.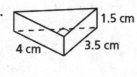

5.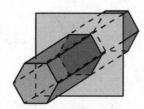

6. Describe the shape formed by the intersection of the plane and the solid.

7. The vertices of $\triangle ABC$ are $A(3, 2)$, $B(6, -1)$, and $C(-1, -3)$. Translate $\triangle ABC$ using the vector $\langle -2, 3 \rangle$. Graph $\triangle ABC$ and its image.

8. The diagram shows the location of a campsite and a trail. You want to choose a campsite that is at least 75 feet from the trail. Does this campsite meet your requirement? Explain.

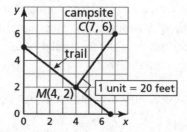

9. Let p be "it is Saturday" and let q be "it is the weekend." Write the conditional statement $p \to q$ and the contrapositive $\sim q \to \sim p$ in words. Then decide whether each statement is true or false.

12.3 Self-Assessment

Use the scale to rate your understanding of the learning target and the success criteria.

1 I do not understand. 2 I can do it with help. 3 I can do it on my own. 4 I can teach someone else.

	Rating	Date
12.3 Volumes of Pyramids		
Learning Target: Find and use volumes of pyramids.	1 2 3 4	
I can find volumes of pyramids.	1 2 3 4	
I can use volumes of pyramids to find measures.	1 2 3 4	
I can find volumes of composite solids containing pyramids.	1 2 3 4	
I can find volumes of similar pyramids.	1 2 3 4	

12.4 Extra Practice

In Exercises 1 and 2, find the surface area of the right cone.

1.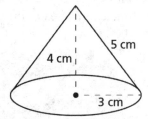

2. A right cone has a diameter of 1.8 inches and a height of 3 inches.

In Exercises 3 and 4, find the volume of the right cone.

3.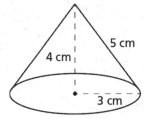

4. A right cone has a radius of 5 feet and a slant height of 13 feet.

5. A right cone has a volume of 528 cubic meters and a diameter of 12 meters. Find its height and slant height.

6. Cone A and cone B are similar. The radius of cone A is 4 centimeters and the radius of cone B is 10 centimeters. The volume of cone A is 134 cubic centimeters. Find the volume of cone B.

7. Find the volume of the composite solid.

   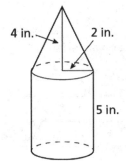

8. A snack stand serves flavored shaved ice in cone-shaped containers and cylindrical containers. Which container gives you more shaved ice for your money? Explain.

   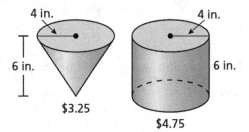

9. A cone has height h and a base with radius r. You want to change the cone so its volume is tripled. What is the new height if you change only the height? What is the new radius if you change only the radius? Explain.

Name _____ Date _____

12.4 Review & Refresh

In Exercises 1 and 2, find the indicated measure.

1. area of a circle with a radius of 8 meters

2. diameter of a circle with an area of 324π square inches

3. A right cone has a radius of 6 feet and a height of 17 feet. Find the surface area of the cone.

4. Find the volume of the cylinder.

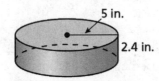

5. Two polygons are similar. The perimeter of one polygon is 78 feet. The ratio of corresponding side lengths is $\frac{5}{3}$. Find two possible perimeters of the other polygon.

6. You cut canned cranberry sauce parallel to its bases. Find the perimeter and area of the cross section formed by the cut.

7. Find the volume of the pyramid.

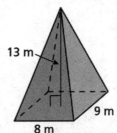

8. Tell whether $\widehat{XY} \cong \widehat{YZ}$. Explain why or why not.

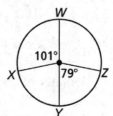

12.4 Self-Assessment

Use the scale to rate your understanding of the learning target and the success criteria.

1 I do not understand. **2** I can do it with help. **3** I can do it on my own. **4** I can teach someone else.

	Rating	Date
12.4 Surface Areas and Volumes of Cones		
Learning Target: Find and use surface areas and volumes of cones.	1 2 3 4	
I can find surface areas of cones.	1 2 3 4	
I can find volumes of cones.	1 2 3 4	
I can find the volumes of composite solids containing cones.	1 2 3 4	
I can find the volumes of similar cones.	1 2 3 4	

Name_____ Date_____

12.5 Extra Practice

In Exercises 1–3, find the surface area of the sphere.

1.
2.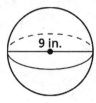
3.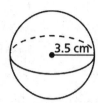

In Exercises 4 and 5, find the indicated measure.

4. Find the radius of a sphere with a surface area of 324π square centimeters.

5. Find the diameter of a sphere with a surface area of 2704π square yards.

In Exercises 6–8, find the volume of the sphere.

6.
7.
8.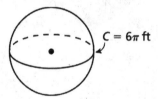

In Exercises 9 and 10, find the surface area and volume of the hemisphere.

9.
10.

In Exercises 11 and 12, find the volume of the sphere with the given surface area.

11. Surface area = 256π in.²
12. Surface area = 400π ft²

In Exercises 13 and 14, find the volume of the composite solid.

13.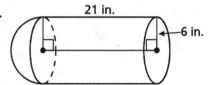
14.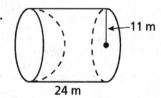

15. A sphere is inscribed in a cube that has a volume of 8 cubic yards. What is the surface area of the sphere? Explain.

Name _____ Date _____

12.5 Review & Refresh

1. Solve the triangle.

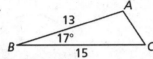

2. The diagram shows the portion of Earth visible to a camera on a weather balloon about 63 miles above Earth at point B. Earth's radius is approximately 4000 miles. Find $m\widehat{AC}$.

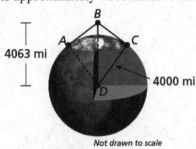

Not drawn to scale

3. The cylinders are similar. Find the volume of cylinder B.

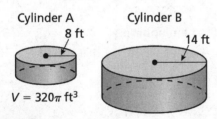

4. Find the missing dimension of the cylinder.

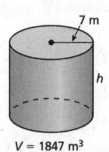

$V = 1847 \text{ m}^3$

5. In rectangle $ABCD$, $AC = 4x + 15$ and $BD = 7x - 9$. Find the lengths of the diagonals of $ABCD$.

6. Determine whether $\overline{RT} \parallel \overline{QU}$. Explain.

In Exercises 7 and 8, find the surface area and the volume of the solid.

7.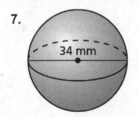

8.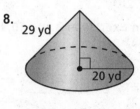

12.5 Self-Assessment

Use the scale to rate your understanding of the learning target and the success criteria.

| 1 I do not understand. | 2 I can do it with help. | 3 I can do it on my own. | 4 I can teach someone else. |

	Rating	Date
12.5 Surface Areas and Volumes of Spheres		
Learning Target: Find and use surface areas and volumes of spheres.	1 2 3 4	
I can find surface areas of spheres.	1 2 3 4	
I can find volumes of spheres.	1 2 3 4	
I can find volumes of composite solids.	1 2 3 4	

12.6 Extra Practice

1. The diagram shows the dimensions of a jar of peanut butter. Peanut butter has a density of about 1.08 grams per cubic centimeter. Find the mass of the peanut butter in the jar.

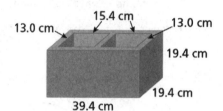

2. A sugar cube has a mass of 2.3 grams. Sugar has a density of about 1.59 grams per cubic centimeter. Find the side length of the sugar cube.

3. The diagram shows the dimensions of a concrete block. Concrete has a density of about 2.24 grams per cubic centimeter. Find the mass of the concrete block.

4. A sheet of aluminum foil is 30.4 centimeters by 25.0 centimeters. The sheet has a mass of 3.28 grams. Aluminum has a density of about 2.7 grams per cubic centimeter. Estimate the thickness of the sheet of aluminum foil.

5. Sand is falling off a conveyor and forms a conical pile. The diameter of the pile is 10 meters and the height is 2 meters.

   a. Sand has a density of about 1631 kilograms per cubic meter. Find the mass of the pile of sand.

   b. The height of the pile increases by 0.1 meter per hour and the radius increases by 0.25 meter per hour. How many cubic meters of sand is added to the pile after 1 hour?

   c. If sand is added to the pile at the same rate for each of the next four hours, will the amount of sand accumulated each hour be the same? Explain.

6. The density of copper is approximately 8.96 grams per cubic centimeter. A hollow copper pipe with a mass of 429 grams has a radius of 2.5 centimeters and a height of 35 centimeters. Estimate the thickness of the pipe.

Name _____ Date _____

12.6 Review & Refresh

1. Show that a quadrilateral with vertices $W(1,12)$, $X(7,2)$, $Y(6,-7)$, and $Z(-6,13)$ is a trapezoid. Then decide whether it is isosceles.

2. Steel has a density of about 7.8 grams per cubic centimeter. A steel rod has a diameter of $\frac{7}{16}$ inch and a length of 9 inches. Find the mass of the steel rod.

3. Find the surface area and the volume of the sphere.

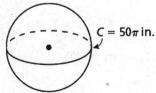

4. Find the volume of the composite solid.

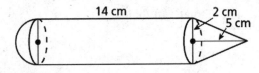

5. The pyramids are similar. Find the volume of pyramid B.

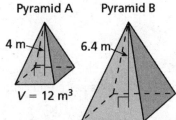

6. Decide whether enough information is given to prove that $\triangle PQS$ and $\triangle RQS$ are congruent. Explain.

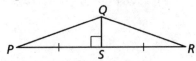

In Exercises 7 and 8, find the value of x. Write your answer in simplest form.

7. 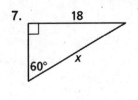   8.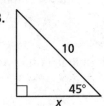

12.6 Self-Assessment

Use the scale to rate your understanding of the learning target and the success criteria.

| 1 I do not understand. | 2 I can do it with help. | 3 I can do it on my own. | 4 I can teach someone else. |

	Rating	Date
12.6 Modeling with Surface Area and Volume		
Learning Target: Understand the concept of density and modeling with volume.	1 2 3 4	
I can explain what density means.	1 2 3 4	
I can use the formula for density to solve problems.	1 2 3 4	
I can use geometric shapes to model objects.	1 2 3 4	
I can solve modeling problems.	1 2 3 4	

Name_____ Date_____

12.7 Extra Practice

In Exercises 1–3, sketch the solid produced by rotating the figure around the given axis. Then identify and describe the solid.

1.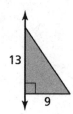
2.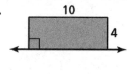
3.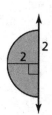

In Exercises 4–6, sketch a two-dimensional shape and an axis of revolution that forms the object shown.

4.
5.
6.

In Exercises 7–10, sketch and describe the solid produced by rotating the figure around the given axis. Then find its surface area and volume.

7.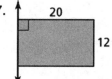
8.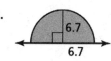

9.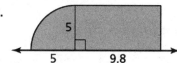
10.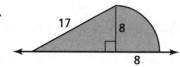

In Exercises 11 and 12, sketch and describe the solid that is produced when the region enclosed by the given equations is rotated around the given axis. Then find the volume of the solid.

11. $x = 0$, $y = 0$, $y = -2x + 7$; x-axis

12. $x = 4$, $y = 0$, $y = 3x$; y-axis

Name _____ Date _____

12.7 Review & Refresh

1. A circular region has a population of about 3.1 million people and a population density of about 9867 people per square mile. Find the radius of the region.

2. Sketch and describe the solid produced by rotating the figure around the given axis. Then find its surface area and volume.

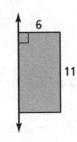

3. The diagram shows dimensions of a cork. Cork has a density of 0.24 gram per cubic centimeter. Find the mass of the cork.

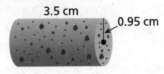

4. Find the value of x. Tell whether the side lengths form a Pythagorean triple.

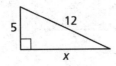

5. Find the surface area of the sphere.

6. Find the volume of the cone.

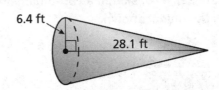

7. You are running on a circular path at a constant rate of 7.4 feet per second. The path is 0.75 mile in diameter. How long will it take you to run three complete laps?

8. Write an equation of a parabola with focus $F(-4, 3)$ and directrix $x = 2$.

12.7 Self-Assessment

Use the scale to rate your understanding of the learning target and the success criteria.

| 1 I do not understand. | 2 I can do it with help. | 3 I can do it on my own. | 4 I can teach someone else. |

	Rating	Date
12.7 Solids of Revolution		
Learning Target: Sketch and use solids of revolution.	1 2 3 4	
I can sketch and describe solids of revolution.	1 2 3 4	
I can find surface areas and volumes of solids of revolution.	1 2 3 4	
I can form solids of revolution in the coordinate plane.	1 2 3 4	

Name_____ Date_____

Chapter 12 Chapter Self-Assessment

Use the scale to rate your understanding of the learning target and the success criteria.

1 I do not understand. **2** I can do it with help. **3** I can do it on my own. **4** I can teach someone else.

	Rating	Date
Chapter 12 Surface Area and Volume		
Learning Target: Understand surface area and volume.	1 2 3 4	
I can describe attributes of solids.	1 2 3 4	
I can find surface areas and volumes of solids.	1 2 3 4	
I can find missing dimensions of solids.	1 2 3 4	
I can solve real-life problems involving surface area and volume.	1 2 3 4	
12.1 Cross Sections of Solids		
Learning Target: Describe and draw cross sections.	1 2 3 4	
I can describe attributes of solids.	1 2 3 4	
I can describe and draw cross sections.	1 2 3 4	
I can solve real-life problems involving cross sections.	1 2 3 4	
12.2 Volumes of Prisms and Cylinders		
Learning Target: Find and use volumes of prisms and cylinders.	1 2 3 4	
I can find volumes of prisms and cylinders.	1 2 3 4	
I can find surface areas and volumes of similar solids.	1 2 3 4	
I can solve real-life problems involving volumes of prisms and cylinders.	1 2 3 4	
12.3 Volumes of Pyramids		
Learning Target: Find and use volumes of pyramids.	1 2 3 4	
I can find volumes of pyramids.	1 2 3 4	
I can use volumes of pyramids to find measures.	1 2 3 4	
I can find volumes of composite solids containing pyramids.	1 2 3 4	
I can find volumes of similar pyramids.	1 2 3 4	

Copyright © Big Ideas Learning, LLC
All rights reserved.

Geometry
Practice Workbook and Test Prep

Chapter 12 Chapter Self-Assessment (continued)

	Rating	Date
12.4 Surface Areas and Volumes of Cones		
Learning Target: Find and use surface areas and volumes of cones.	1 2 3 4	
I can find surface areas of cones.	1 2 3 4	
I can find volumes of cones.	1 2 3 4	
I can find the volumes of composite solids containing cones.	1 2 3 4	
I can find the volumes of similar cones.	1 2 3 4	
12.5 Surface Areas and Volumes of Spheres		
Learning Target: Find and use surface areas and volumes of spheres.	1 2 3 4	
I can find surface areas of spheres.	1 2 3 4	
I can find volumes of spheres.	1 2 3 4	
I can find volumes of composite solids.	1 2 3 4	
12.6 Modeling with Surface Area and Volume		
Learning Target: Understand the concept of density and modeling with volume.	1 2 3 4	
I can explain what density means.	1 2 3 4	
I can use the formula for density to solve problems.	1 2 3 4	
I can use geometric shapes to model objects.	1 2 3 4	
I can solve modeling problems.	1 2 3 4	
12.7 Solids of Revolution		
Learning Target: Sketch and use solids of revolution.	1 2 3 4	
I can sketch and describe solids of revolution.	1 2 3 4	
I can find surface areas and volumes of solids of revolution.	1 2 3 4	
I can form solids of revolution in the coordinate plane.	1 2 3 4	

Name_____ Date_____

Chapter 12 Test Prep

1. What is the volume of the sphere with a surface area of 2304π square millimeters? Round the answer to the nearest tenth.

   mm³

2. A circular region with an 8-mile radius has a population density of 942 people per square mile. How many people live in the region?

   people

3. What solid is produced by rotating the figure around the given axis?

   Ⓐ square pyramid
   Ⓑ cone
   Ⓒ triangular pyramid
   Ⓓ triangular prism

4. What is the volume of the composite solid?

   Ⓐ about 54.8 in.³
   Ⓑ about 73.7 in.³
   Ⓒ about 111.4 in.³
   Ⓓ about 186.8 in.³

   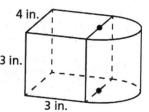

5. Which of the following triangles are similar to $\triangle ABC$? Select all that apply.

   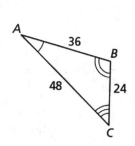

   Ⓐ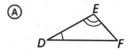

   Ⓑ

   Ⓒ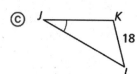

   Ⓓ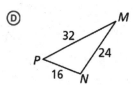

Chapter 12 Test Prep (continued)

6. What is the value of y that makes the quadrilateral a parallelogram?

Ⓐ 9
Ⓑ 10
Ⓒ 20
Ⓓ 24

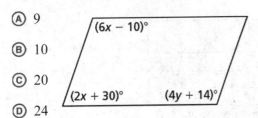

7. The surface area of the cone is 703.7 square meters. What is the height of the cone?

Ⓐ 13.7 m
Ⓑ 14 m
Ⓒ 24 m
Ⓓ 25 m

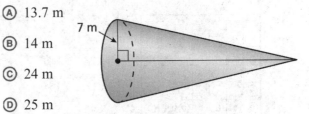

8. Which definitions and properties can you use to complete the proof? Select all that apply.

Ⓐ Subtraction Property of Equality
Ⓑ Transitive Property of Congruence
Ⓒ Division Property of Equality
Ⓓ Symmetric Property of Congruence
Ⓔ Substitution Property of Equality
Ⓕ Definition of segment bisector

Given $QS = TV$, $\overline{RU}$ bisects $\overline{QS}$ and $\overline{TV}$
Prove $RS = TU$

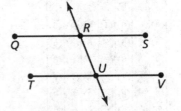

STATEMENTS	REASONS
1. $QS = TV$, $\overline{RU}$ bisects $\overline{QS}$ and $\overline{TV}$	1. Given
2. $\overline{QR} \cong \overline{RS}$, $\overline{TU} \cong \overline{UV}$	2. _____
3. $QR = RS$, $TU = UV$	3. Definition of congruent segments
4. $QS = QR + RS$ $TV = TU + UV$	4. Segment Addition Postulate
5. $QR + RS = TU + UV$	5. _____
6. $RS + RS = TU + TU$	6. _____
7. $2RS = 2TU$	7. Distributive Property
8. $RS = TU$	8. _____

9. What is the surface area of the hemisphere?

Ⓐ about 265.5 in.²
Ⓑ about 398.2 in.²
Ⓒ about 1061.9 in.²
Ⓓ about 1592.8 in.²

10. If the segment lengths 8, 11, and 16 form a triangle, is the triangle *acute*, *right*, or *obtuse*?

Ⓐ acute
Ⓑ right
Ⓒ obtuse
Ⓓ The segment lengths do not form a triangle.

Name_____ Date_____

Chapter 12 Test Prep (continued)

11. Which of the following are polyhedra? Select all that apply.

Ⓐ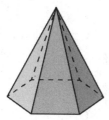

Ⓑ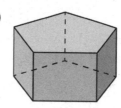

Ⓒ

Ⓓ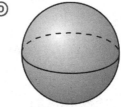

12. The volume of the cone is 152.1 cubic centimeters. What is the surface area of the cone?

   Ⓐ about 221.2 cm²

   Ⓑ about 561.1 cm²

   Ⓒ about 763.7 cm²

   Ⓓ about 974.5 cm²

   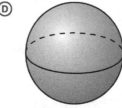
   11 cm

13. A pyramid with a square base has a volume of 96 cubic feet and a height of 8 feet. What is the side length of the base?

   Ⓐ $2\sqrt{3}$ ft

   Ⓑ 6 ft

   Ⓒ 18 ft

   Ⓓ 36 ft

14. The pyramids are similar. What is the volume of pyramid B?

   Ⓐ 122.88 in.³

   Ⓑ 153.6 in.³

   Ⓒ 192 in.³

   Ⓓ 300 in.³

   Pyramid A: 15 in., V = 240 in.³ Pyramid B: 12 in.

15. Describe the shape formed by the intersection of the plane and the solid.

   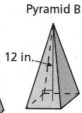

   _____

Copyright © Big Ideas Learning, LLC
All rights reserved.

Name _____ Date _____

Chapter 12 Test Prep (continued)

16. What is $m\widehat{AE}$?

  Ⓐ 18°
  Ⓑ 54°
  Ⓒ 72°
  Ⓓ 108°

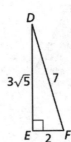

17. What is tan F?

  Ⓐ $\dfrac{7}{2}$
  Ⓑ $\dfrac{3\sqrt{5}}{2}$
  Ⓒ $\dfrac{3\sqrt{5}}{7}$
  Ⓓ $\dfrac{2\sqrt{5}}{15}$

18. What is the value of x that makes $m \parallel n$?

  Ⓐ 9.25
  Ⓑ 13.25
  Ⓒ 14
  Ⓓ 16

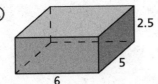

19. What are the possible values of x?

  Ⓐ $0 < x < 6$
  Ⓑ $\dfrac{7}{3} < x < 6$
  Ⓒ $x > 6$
  Ⓓ $x = 6$

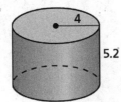

20. Which of the following solids has the greatest volume?

  Ⓐ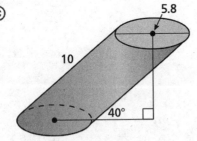
  Ⓑ
  Ⓒ
  Ⓓ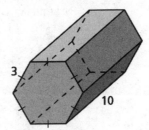

21. The diagram shows the dimensions of a bar of silver. Silver has a density of about 10.49 grams per cubic centimeter. What is the mass of the bar of silver?

  Ⓐ 1.7 kg
  Ⓑ 15.7 kg
  Ⓒ 63.6 kg
  Ⓓ 1730.9 kg

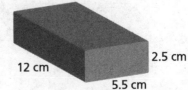

13.1 Extra Practice

In Exercises 1 and 2, find the number of possible outcomes in the sample space. Then list the possible outcomes.

1. A stack of cards contains the thirteen clubs from a standard deck of cards. You pick a card from the stack and flip two coins.

2. You spin a spinner with 5 equal sections (one blue, one green, one red, one yellow, and one purple) and roll a die.

3. There are two bags, each containing 10 tiles numbered 1 through 10. When one tile is chosen from each bag, there are 100 possible outcomes. Find the probability that (a) the sum of the two numbers is *not* 10 and (b) the product of the numbers is greater than 10.

4. You throw a dart at the board shown. Your dart is equally likely to hit any point inside the square board. What is the probability your dart lands in the black circle?

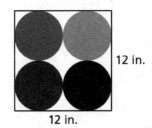

5. The sections of a spinner are numbered 1 through 12. Each section of the spinner has the same area. You spin the spinner 180 times. The table shows the results. For which number is the experimental probability of stopping on the number the same as the theoretical probability?

Spinner Results											
1	2	3	4	5	6	7	8	9	10	11	12
13	21	22	20	11	8	14	9	15	12	18	17

6. A manufacturer measures the length of 1500 nails and finds 100 of them to be too short. Predict the number of short nails in a box containing 2400 nails. Explain your reasoning.

Name _____ Date _____

13.1 Review & Refresh

1. Sketch the solid produced by rotating the figure around the given axis. Then identify and describe the solid.

   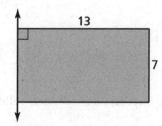

2. A logger cuts down a pine tree and removes its limbs. The remaining trunk is 50 feet long and has a diameter of 1.8 feet. The density of pine is 0.5 gram per cubic centimeter. Find the mass of the tree trunk.

3. Find the surface area and volume of the hemisphere.

   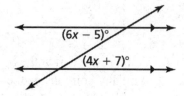

4. The spinner is divided into sections with the same area. You spin the spinner 20 times. It stops on an odd number 9 times. Compare the experimental probability of spinning an odd number with the theoretical probability.

5. Find the value of x that makes $\triangle DEF \sim \triangle PQR$.

   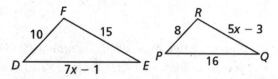

6. Find the value of x.

   (6x − 5)°
   (4x + 7)°

13.1 Self-Assessment

Use the scale to rate your understanding of the learning target and the success criteria.

| 1 I do not understand. | 2 I can do it with help. | 3 I can do it on my own. | 4 I can teach someone else. |

	Rating	Date
13.1 Sample Spaces and Probability		
Learning Target: Find sample spaces and probabilities of events.	1 2 3 4	
I can list the possible outcomes in a sample space.	1 2 3 4	
I can find theoretical probabilities.	1 2 3 4	
I can find experimental probabilities.	1 2 3 4	

13.2 Extra Practice

1. Complete the two-way table.

		Arrival		
		Tardy	On Time	Total
Method	Walk	22		
	City Bus			60
	Total		58	130

2. A survey was taken of 100 families with one child and 86 families with two or more children to determine whether they were saving for college. Of those, 94 of the families with one child and 60 of the families with two or more children were saving for college. Organize these results in a two-way table. Then find and interpret the marginal frequencies.

3. In a survey, 214 ninth graders played video games every day of the week and 22 ninth graders did not play video games every day of the week. Of those that played every day of the week, 36 had trouble sleeping at night. Of those that did not play every day of the week, 7 had trouble sleeping at night. Make a two-way table that shows the joint and marginal relative frequencies. Interpret one of the joint relative frequencies and one of the marginal relative frequencies.

4. Use the survey results from Exercise 2 to make a two-way table that shows the conditional relative frequencies based on (a) the row totals and (b) the column totals. Interpret one of the conditional relative frequencies in each table.

13.2 Review & Refresh

1. A survey finds that 35 people go to the gym regularly and 87 people do not go to the gym regularly. Of those who go regularly, 9 people do not sleep well. Of those who do not go regularly, 36 people do not sleep well. Make a two-way table that shows the conditional relative frequencies based on the gym attendance totals. Then interpret one of the conditional relative frequencies.

2. When four coins are flipped, there are 16 possible outcomes. Find the probability that there are *not* exactly three tails.

3. Sketch and describe the solid produced by rotating the figure around the given axis. Then find its surface area and volume.

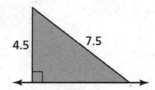

4. Find $m\angle BAC$.

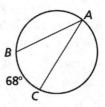

5. Find the values of x and y that make the quadrilateral a parallelogram.

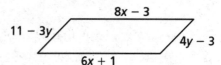

13.2 Self-Assessment

Use the scale to rate your understanding of the learning target and the success criteria.

| 1 I do not understand. | 2 I can do it with help. | 3 I can do it on my own. | 4 I can teach someone else. |

	Rating	Date
13.2 Two-Way Tables and Probability		
Learning Target: Use two-way tables to represent data and find probabilities.	1 2 3 4	
I can make two-way tables.	1 2 3 4	
I can find and interpret relative frequencies and conditional relative frequencies.	1 2 3 4	
I can use conditional relative frequencies to find probabilities.	1 2 3 4	

13.3 Extra Practice

1. You have three $5 bills and four $10 bills in your wallet. You randomly select two different dollar bills from your wallet. Find the probability that you select a $5 bill second given that you randomly selected a $10 bill first.

2. The two-way table shows the number of dogs and cats at three local animal shelters. Find the probability of each.

		Shelter		
		A	B	C
Animal	Dog	10	18	12
	Cat	24	20	35

   a. $P(\text{Shelter A} \mid \text{dog})$

   b. $P(\text{cat} \mid \text{Shelter B})$

3. Find the probability in Exercise 2(a) using the formula for conditional probability.

4. At a farmer's market, 75% of customers buy produce, 6% of customers buy produce and meat, and 1% of customers buy produce and flowers.

   a. What is the probability that a customer who buys produce also buys meat?

   b. What is the probability that a customer who buys produce also buys flowers?

5. You want to find the quickest route to work. You map out three routes. Before work, you randomly select a route and record whether you were late or on time. The table shows your findings. Assuming you leave at the same time each morning, which route should you use? Explain.

Route	On Time	Late
A	ʜʜʜ ʜʜʜ ʜʜʜ	III
B	ʜʜʜ ʜʜʜ III	IIII
C	ʜʜʜ ʜʜʜ ʜʜʜ I	ʜʜʜ

6. In a survey, 62% of respondents have a tablet, 30% have a laptop, and 12% of the respondents who have a laptop also have a tablet.

   a. What is the probability that a person from the survey has both a laptop and tablet?

   b. What is the probability that a person from the survey who has a tablet also has a laptop?

Name _____ Date _____

13.3 Review & Refresh

1. Use the data to create a two-way table that shows the joint and marginal relative frequencies.

		Takes Advanced Courses		
		Yes	No	Total
Gender	Female	63	112	175
	Male	54	107	161
	Total	117	219	336

2. Find the area of a circle with a diameter of 9 centimeters.

3. Find the value of x.

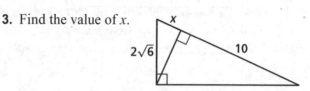

4. Sketch and describe the solid that is produced when the region enclosed by the graphs of the equations $x = -2$, $y = 0$, and $y = -3x$ is rotated around the y-axis. Then find the volume of the solid.

5. The two-way table shows the numbers of people who wear glasses and contacts. Find each probability.

		Glasses	
		Yes	No
Contacts	Yes	48	21
	No	24	33

a. $P(\text{glasses} \mid \text{contacts})$

b. $P(\text{contacts} \mid \text{no glasses})$

13.3 Self-Assessment

Use the scale to rate your understanding of the learning target and the success criteria.

| 1 I do not understand. | 2 I can do it with help. | 3 I can do it on my own. | 4 I can teach someone else. |

	Rating	Date
13.3 Conditional Probability		
Learning Target: Find and use conditional probabilities.	1 2 3 4	
I can explain the meaning of conditional probability.	1 2 3 4	
I can find conditional probabilities.	1 2 3 4	
I can make decisions using probabilities.	1 2 3 4	

13.4 Extra Practice

1. You have three white golf balls and two yellow golf balls in a bag. You randomly select one golf ball to hit now and another golf ball to place in your pocket. Use a sample space to determine whether randomly selecting a white golf ball first and then a white golf ball second are independent events.

2. You roll a six-sided die two times. Use a conditional probability to determine whether getting a 1 each time are independent events.

3. A principal surveys a random sample of high school students in three grades. The survey asks whether students plan to attend the homecoming football game. The results, given as joint relative frequencies, are shown in the two-way table. Determine whether attending the game and being a junior are independent events.

		Grade		
		Sophomore	Junior	Senior
Response	Yes	0.223	0.251	0.342
	No	0.033	0.102	0.049

4. A spinner is divided into equal parts. Find the probability you get a 8 on your first spin and a number less than 6 on your second spin.

5. A music streaming queue shows 16 R&B songs and 12 pop songs. You randomly choose two songs to listen to. Find the probability that both events A and B will occur.

   Event A: The first song is a pop song.
   Event B: The second song is a R&B song.

6. You randomly select three cards from a standard deck of 52 playing cards. What is the probability that all three cards are hearts when (a) you replace each card before selecting the next card, and (b) you do not replace each card before selecting the next card?

Name _____ Date _____

13.4 Review & Refresh

1. You flip a coin and roll a six-sided die. Find the probability that you get tails when flipping the coin and a multiple of 3 when rolling the die.

2. Write an equation of the parabola with focus $F(-2, 5)$ and directrix $y = -1$.

3. Find the measures of the numbered angles in rhombus $WXYZ$.

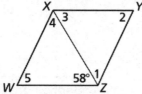

4. Find the coordinates of the centroid of $\triangle ABC$ with vertices $A(3, -5)$, $B(-6, 4)$, and $C(0, 1)$.

5. You randomly draw a marble out of a bag containing 7 purple marbles, 9 yellow marbles, 3 orange marbles, and 6 green marbles. Find the probability of drawing a marble that is *not* green.

6. Find the distance from the point $A(-7, -3)$ to the line $y = 4x - 3$.

7. Find the value of x.

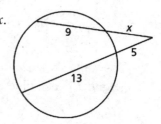

8. A survey asks 46 females and 61 males whether they use a journal. Complete the two-way table. Then interpret the marginal frequencies.

		Journal		
		Yes	No	Total
Gender	Female		22	
	Male	25		
	Total			

13.4 Self-Assessment

Use the scale to rate your understanding of the learning target and the success criteria.

| 1 I do not understand. | 2 I can do it with help. | 3 I can do it on my own. | 4 I can teach someone else. |

	Rating	Date
13.4 Independent and Dependent Events		
Learning Target: Understand and find probabilities of independent and dependent events.	1 2 3 4	
I can explain how independent events and dependent events are different.	1 2 3 4	
I can determine whether events are independent.	1 2 3 4	
I can find probabilities of independent and dependent events.	1 2 3 4	

224 Geometry
Practice Workbook and Test Prep

13.5 Extra Practice

In Exercises 1 and 2, events A and B are disjoint. Find P(A or B).

1. $P(A) = 0.3, P(B) = 0.55$

2. $P(A) = \frac{2}{3}, P(B) = \frac{1}{6}$

In Exercises 3–6, a ticket is randomly chosen from a bag. The bag contains tickets numbered 1 to 16. Find the probability of the event.

3. choosing an odd number *or* a multiple of 4

4. choosing an even number *or* a factor of 25

5. choosing a number less than or equal to 10 *or* a multiple of 3

6. choosing a number greater than 3 *or* a perfect square

7. You survey 135 students. Of the students in the survey, 81 have a movie streaming subscription, 54 have a music streaming subscription, and 26 have both a movie steaming subscription and a music streaming subscription. What is the probability that a student in the survey has a movie streaming subscription *or* a music streaming subscription?

8. Out of 120 student parents, 90 of them can chaperone the homecoming dance or the prom. There are 40 parents who can chaperone the homecoming dance and 65 parents who can chaperone the prom. What is the probability that a randomly selected parent can chaperone both the homecoming dance *and* the prom?

9. A football team scores a touchdown first 75% of the time when they start with the ball. The team does not score first 51% of the time when their opponent starts with the ball. The team who gets the ball first is determined by a coin toss. What is the probability that the team scores a touchdown first?

13.5 Review & Refresh

1. You randomly draw a card from a standard deck of 52 playing cards, set it aside, and then randomly draw another card from the deck. Find the probability that both cards are face cards.

2. Find the value of each variable using sine and cosine.

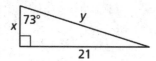

3. You eat one cup of ice cream for dessert each day. You buy the full carton of ice cream shown. How many days can you have a cup of ice cream? (1 cup ≈ 14.4 in.^{3})

In Exercises 4 and 5, find the missing probability.

4. $P(A) = 0.3$
   $P(A \text{ and } B) = 0.18$
   $P(B \mid A) =$ ____

5. $P(A) = 0.56$
   $P(B) = 0.28$
   $P(A \text{ and } B) = 0.32$
   $P(B \text{ or } A) =$ ____

In Exercises 6–8, find the measure of the arc where $\overline{RT}$ is a diameter.

6. $\widehat{QR}$

7. $\widehat{QS}$

8. $\widehat{QRS}$

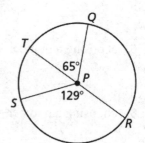

13.5 Self-Assessment

Use the scale to rate your understanding of the learning target and the success criteria.

| 1 I do not understand. | 2 I can do it with help. | 3 I can do it on my own. | 4 I can teach someone else. |

	Rating	Date
13.5 Probability of Disjoint and Overlapping Events		
Learning Target: Find probabilities of disjoint and overlapping events.	1 2 3 4	
I can explain how disjoint events and overlapping events are different.	1 2 3 4	
I can find probabilities of disjoint events.	1 2 3 4	
I can find probabilities of overlapping events.	1 2 3 4	
I can solve real-life problems using more than one probability rule.	1 2 3 4	

Name_____ Date_____

13.6 Extra Practice

In Exercises 1 and 2, find the number of ways you can arrange (a) all of the numbers and (b) 3 of the numbers in the given amount.

1. $2,564,783

2. $4,128,675,309

3. Your rock band has nine songs recorded but you only want to put five of them on your demo CD to hand out to publishers. How many possible ways could the five songs be ordered on your demo CD?

4. A witness at the scene of a hit-and-run accident saw that the car that caused the accident had a license plate with only the letters I, R, L, T, O, and A. Find the probability that the license plate starts with a T and ends with an R.

In Exercises 5 and 6, count the possible combinations of r letters chosen from the given list.

5. G, H, I, J, K; $r = 2$

6. P, Q, R, S, T, U, V, W; $r = 4$

7. How many possible combinations of three colors can be chosen from the seven colors of the rainbow?

8. You are ordering a smoothie with 3 fruits and 2 vegetables. The menu shows the possible choices. How many different smoothies are possible?

9. The organizer of a gift exchange asks each of 5 people to bring 1 gift card from a list of 8 gift cards. Assuming each person randomly chooses a gift card, what is the probability that at least 2 of the 5 people bring the same gift card?

13.6 Review & Refresh

1. There are three grape juice boxes and two apple juice boxes in your refrigerator. You randomly select one juice box to drink now and another juice box to drink with lunch. Use a sample space to determine whether randomly selecting a grape juice box first and a grape juice box second are independent events.

2. Events A and B are dependent. Suppose $P(A \text{ and } B) = 0.08$ and $P(A) = 0.72$. Find $P(B \mid A)$.

3. Events A and B are disjoint. Find $P(A \text{ or } B)$ when $P(A) = 0.35$ and $P(B) = 0.28$.

In Exercises 4 and 5, evaluate the expression.

4. $_8P_5$

5. $_6C_2$

6. A Ferris wheel has a diameter of 350 feet. How far does a passenger car travel when it makes 9 revolutions?

7. Point D is the centroid of $\triangle ABC$, and $CD = 24$. Find CE and DE.

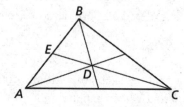

8. Find the measure of the exterior angle.

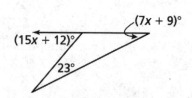

9. Find the dimensions of the rectangular prism.

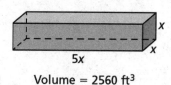

Volume = 2560 ft³

13.6 Self-Assessment

Use the scale to rate your understanding of the learning target and the success criteria.

| 1 I do not understand. | 2 I can do it with help. | 3 I can do it on my own. | 4 I can teach someone else. |

	Rating	Date
13.6 Permutations and Combinations		
Learning Target: Count permutations and combinations.	1 2 3 4	
I can explain the difference between permutations and combinations.	1 2 3 4	
I can find numbers of permutations and combinations.	1 2 3 4	
I can find probabilities using permutations and combinations.	1 2 3 4	

Name_____ Date_____

13.7 Extra Practice

In Exercises 1 and 2, make a table and draw a histogram showing the probability distribution for the random variable.

1. d = the number on a rubber duck randomly chosen from a pool that contains 8 ducks labeled "1," 4 ducks labeled "2," 5 ducks labeled "3," and 3 ducks labeled "4."

2. x = the product when two six-sided dice are rolled.

3. Use the probability distribution to determine (a) the number that is most likely to be spun on a spinner, and (b) the probability of spinning a perfect square.

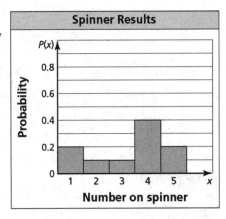

4. According to a survey, 78% of women in a city watch professional football. You ask 4 randomly chosen women from the city whether they watch professional football.

   a. Draw a histogram of the binomial distribution for your survey.

   b. What is the most likely outcome of the survey?

   c. What is the probability that at most one woman watches professional football?

13.7 Review & Refresh

1. Use the probability distribution below to determine the most likely number of appointments on a given day.

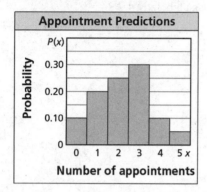

In Exercises 2 and 3, count the possible combinations of *r* letters chosen from the given list.

2. A, B, C, D, E; $r = 2$

3. S, T, U, V, W, X; $r = 4$

4. Find the coordinates of the intersection of the diagonals of $\square DEFG$ with vertices $D(4, 6)$, $E(2, 0)$, $F(-1, 1)$, and $G(1, 7)$.

5. Let $\angle H$ be an acute angle with $\cos H = 0.24$. Use technology to approximate $m\angle H$.

6. You collect data about school breakfast. Of the 59 students who eat breakfast, 26 students get a smoothie, 38 students get a milk carton, and 14 students get both a smoothie and a milk carton. What is the probability that a student gets a smoothie *or* a milk carton?

7. A box contains four red markers, five blue markers, and ten green markers. You randomly draw a marker and set it aside. Then you randomly draw another marker.

   Event *A*: You draw a blue marker first.

   Event *B*: You draw a blue marker second.

   Tell whether the events are independent or dependent. Explain your reasoning.

8. Point *B* is a point of tangency. Find the radius *r* of $\odot C$.

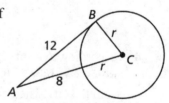

13.7 Self-Assessment

Use the scale to rate your understanding of the learning target and the success criteria.

| 1 I do not understand. | 2 I can do it with help. | 3 I can do it on my own. | 4 I can teach someone else. |

	Rating	Date
13.7 Binomial Distributions		
Learning Target: Understand binomial distributions.	1 2 3 4	
I can explain the meaning of a probability distribution.	1 2 3 4	
I can construct and interpret probability distributions.	1 2 3 4	
I can find probabilities using binomial distributions.	1 2 3 4	

Name_____ Date_____

 Chapter Self-Assessment

Use the scale to rate your understanding of the learning target and the success criteria.

1 I do not understand. **2** I can do it with help. **3** I can do it on my own. **4** I can teach someone else.

	Rating	Date
Chapter 13 Probability		
Learning Target: Understand probability.	1 2 3 4	
I can define theoretical and experimental probability.	1 2 3 4	
I can use two-way tables to find probabilities.	1 2 3 4	
I can compare independent and dependent events.	1 2 3 4	
I can construct and interpret probability and binomial distributions.	1 2 3 4	
13.1 Sample Spaces and Probability		
Learning Target: Find sample spaces and probabilities of events.	1 2 3 4	
I can list the possible outcomes in a sample space.	1 2 3 4	
I can find theoretical probabilities.	1 2 3 4	
I can find experimental probabilities.	1 2 3 4	
13.2 Two-Way Tables and Probability		
Learning Target: Use two-way tables to represent data and find probabilities.	1 2 3 4	
I can make two-way tables.	1 2 3 4	
I can find and interpret relative frequencies and conditional relative frequencies.	1 2 3 4	
I can use conditional relative frequencies to find probabilities.	1 2 3 4	
13.3 Conditional Probability		
Learning Target: Find and use conditional probabilities.	1 2 3 4	
I can explain the meaning of conditional probability.	1 2 3 4	
I can find conditional probabilities.	1 2 3 4	
I can make decisions using probabilities.	1 2 3 4	

Name _____ Date _____

Chapter 13 Chapter Self-Assessment (continued)

	Rating	Date
13.4 Independent and Dependent Events		
Learning Target: Understand and find probabilities of independent and dependent events.	1 2 3 4	
I can explain how independent events and dependent events are different.	1 2 3 4	
I can determine whether events are independent.	1 2 3 4	
I can find probabilities of independent and dependent events.	1 2 3 4	
13.5 Probability of Disjoint and Overlapping Events		
Learning Target: Find probabilities of disjoint and overlapping events.	1 2 3 4	
I can explain how disjoint events and overlapping events are different.	1 2 3 4	
I can find probabilities of disjoint events.	1 2 3 4	
I can find probabilities of overlapping events.	1 2 3 4	
I can solve real-life problems using more than one probability rule.	1 2 3 4	
13.6 Permutations and Combinations		
Learning Target: Count permutations and combinations.	1 2 3 4	
I can explain the difference between permutations and combinations.	1 2 3 4	
I can find numbers of permutations and combinations.	1 2 3 4	
I can find probabilities using permutations and combinations.	1 2 3 4	
13.7 Binomial Distributions		
Learning Target: Understand binomial distributions.	1 2 3 4	
I can explain the meaning of a probability distribution.	1 2 3 4	
I can construct and interpret probability distributions.	1 2 3 4	
I can find probabilities using binomial distributions.	1 2 3 4	

Name_____ Date_____

Chapter 13 Test Prep

1. The diagram shows the dimensions of a stick of butter. Butter has a density of about 0.86 gram per cubic centimeter. What is the mass of the stick of butter?

   Ⓐ about 111.8 g

   Ⓑ about 130.0 g

   Ⓒ about 151.2 g

   Ⓓ about 157.4 g

   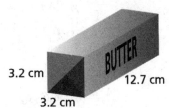

2. Which of the following is an equation of the line passing through the point $P(3, -7)$ that is parallel to the line $y = -\frac{2}{3}x + 4$?

   Ⓐ $y = -\frac{2}{3}x - 7$

   Ⓑ $y = -\frac{2}{3}x - 5$

   Ⓒ $y = \frac{3}{2}x - 7$

   Ⓓ $y = \frac{3}{2}x - \frac{23}{2}$

3. You randomly draw a card from a standard deck of 52 playing cards. What is the probability that you draw an ace *or* a diamond?

   Ⓐ about 1.4%

   Ⓑ about 1.9%

   Ⓒ about 30.8%

   Ⓓ about 32.7%

4. You and a friend participate in a gift exchange. Each participant will draw and keep a number from a hat. There are 16 pieces of paper numbered 1 to 16 in the hat. You draw first and then your friend draws second. What is the probability that you draw a number greater than 12 and then your friend draws a number greater than 10?

   Ⓐ $\dfrac{1}{256}$

   Ⓑ $\dfrac{5}{64}$

   Ⓒ $\dfrac{1}{12}$

   Ⓓ $\dfrac{7}{12}$

5. A six-sided die is rolled twice and a coin is flipped. What is the number of possible outcomes in the sample space?

   _____

Chapter 13 Test Prep (continued)

6. What is the solution of $_nC_{n-2} = {_nC_{n-4}}$?

  Ⓐ $n = -1$
  Ⓑ $n = 6$
  Ⓒ $n = 8$
  Ⓓ $n = 12$

7. What is the area of the kite?

  Ⓐ 240.5 cm^2
  Ⓑ 480 cm^2
  Ⓒ 481 cm^2
  Ⓓ 960 cm^2

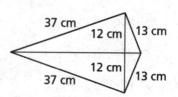

8. What is an equation of the perpendicular bisector of $\overline{AB}$ with endpoints $A(-3, 8)$ and $B(9, -2)$?

_____

9. The figure has x lines of symmetry. What is the value of $x^2 + 2x - 1$?

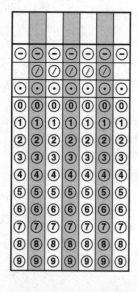

10. Let $P(A) = 0.4$, $P(B) = 0.6$, and $P(A \text{ and } B) = 0.12$. What is $P(A|B)$?

$P(A|B) = $

11. A survey asks students whether they own a laptop. The two-way table shows the results. What does 49 represent in the table?

  Ⓐ the number of males who own a laptop
  Ⓑ the number of females who do not own a laptop
  Ⓒ the number of students who own a laptop
  Ⓓ the number of females who own a laptop

		Own a Laptop		
		Yes	No	Total
Gender	Male	58	79	137
	Female	49	60	109
	Total	107	139	246

Chapter 13 Test Prep (continued)

12. Fifteen students apply to be an officer for a club. The positions are president, vice president, treasurer, and secretary. How many outcomes are possible?

13. Which of the following is a chord in the diagram? Select all that apply.

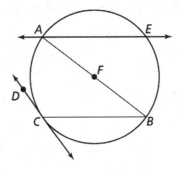

- Ⓐ $\overline{CD}$
- Ⓑ $\overline{BC}$
- Ⓒ $\overline{AE}$
- Ⓓ $\overline{BF}$
- Ⓔ $\overline{AB}$

14. A company surveys all of its employees about which benefit package they want. The two-way table shows the results. Which statement about the two-way table is false?

- Ⓐ The company has 20 employees.
- Ⓑ The probability that a randomly selected employee from the survey chooses Package A is 45%.
- Ⓒ Five employees choose Package A and three employees choose Package B.
- Ⓓ The probability that a randomly selected employee from the survey is male is 40%.

		Benefit Package	
		A	B
Gender	Male	5	3
	Female	4	8

15. A bag contains 5 marbles that are each a different color. A marble is drawn, its color is recorded, and then the marble is put back in the bag. The table shows the results after 50 draws. For which marble(s) is the experimental probability of drawing the marble less than the theoretical probability? Select all that apply.

- Ⓐ green
- Ⓑ yellow
- Ⓒ black
- Ⓓ pink
- Ⓔ white

Drawing Results				
green	yellow	black	pink	white
8	13	10	4	15

Chapter 13 Test Prep (continued)

16. There are seven songs on your playlist: four pop songs and three rock songs. You randomize the playlist. What is the probability that the second song is a rock song given that the first song is a pop song?

Ⓐ $\dfrac{1}{12}$ Ⓑ $\dfrac{1}{7}$

Ⓒ $\dfrac{2}{7}$ Ⓓ $\dfrac{1}{2}$

17. What is the value of x?

Ⓐ 14
Ⓑ 16
Ⓒ 30
Ⓓ 46

18. What is the volume of the prism?

Ⓐ 72 in.^{3}
Ⓑ $24\sqrt{13}$ in.^{3}
Ⓒ $24\sqrt{17}$ in.^{3}
Ⓓ 108 in.^{3}

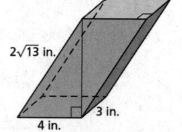

19. A sphere has a diameter of $5x + 4$ inches and a volume of 2304π cubic inches. What is the value of x?

Ⓐ 1.6
Ⓑ 4
Ⓒ 8.8
Ⓓ 12

20. A solid is produced when the region enclosed by $x = 2$, $y = 0$, and $y = \dfrac{3}{2}x$ is rotated around the y-axis. What is the volume of the solid?

Ⓐ 4π cubic units
Ⓑ 6π cubic units
Ⓒ 8π cubic units
Ⓓ 12π cubic units

21. At a clothing store, 44% of customers buy a shirt, 23% of customers buy a pair of sneakers, and 15% of customers buy a shirt and a pair of sneakers. What is the probability that a customer who buys a pair of sneakers also buys a shirt?

Ⓐ about 10.1%
Ⓑ about 34.1%
Ⓒ about 52.3%
Ⓓ about 65.2%

Name_____   Date_____

Geometry Post-Course Test

1. Which congruence transformation maps $\triangle XYZ$ to $\triangle X''Y''Z''$? Select all that apply.

   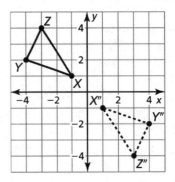

   Ⓐ reflection in the y-axis, followed by a reflection in the x-axis

   Ⓑ rotation of 90° about the origin, followed by a reflection in the y-axis

   Ⓒ rotation of 270° about the origin, followed by a reflection in the x-axis

   Ⓓ reflection in the x-axis, followed by a reflection in the y-axis

   Ⓔ reflection in the x-axis, followed by a rotation of 90° about the origin

2. What is the perimeter of the parallelogram?

   Ⓐ 12 units
   Ⓑ 15 units
   Ⓒ 30 units
   Ⓓ 44 units

   (parallelogram with sides $5y-9$, $x+2$, $2y+3$, $2x$)

3. What is $m\overset{\frown}{BD}$?

   Ⓐ 11°
   Ⓑ 18°
   Ⓒ 24°
   Ⓓ 32°

   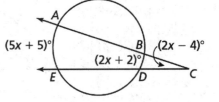

4. Points B and D are points of tangency. What are the values of x? Select all that apply.

   Ⓐ −3
   Ⓑ −2
   Ⓒ 2
   Ⓓ 3
   Ⓔ 6
   Ⓕ 16

   (circle C with tangent segments $2x^2 + 4x$ and $-x^2 + x + 18$ from A)

5. What is another name for $\overleftrightarrow{CD}$? Select all that apply.

   Ⓐ $\overrightarrow{DC}$
   Ⓑ $\overrightarrow{AB}$
   Ⓒ line m
   Ⓓ $\overline{CE}$
   Ⓔ line n
   Ⓕ $\overline{ED}$

   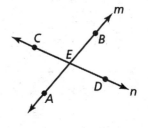

6. What is an equation of the perpendicular bisector of the segment with endpoints $M(-3, 8)$ and $N(1, -2)$?

_____

Geometry Post-Course Test (continued)

7. What is the area of the trapezoid?

   Ⓐ 20 units²

   Ⓑ $4\sqrt{5} + 4\sqrt{10}$ units²

   Ⓒ $20\sqrt{2}$ units²

   Ⓓ 40 units²

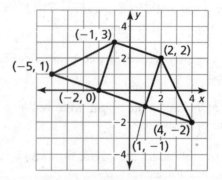

8. The triangles are similar. What is the area of triangle A?

9. What is the area of the shaded region? Round to the nearest hundredth.

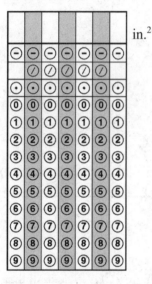   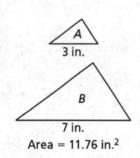   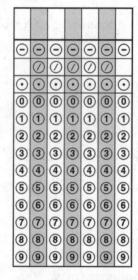   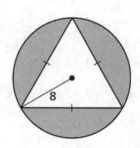

10. △LMN has vertices L(1, 0), M(0, 3) and N(2, 2). Which graph shows the image of △LMN after a dilation with a factor of 3, followed by a translation 4 units down and 2 units left?

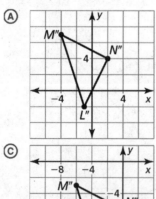   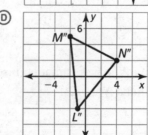

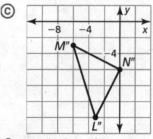

Name_____ Date_____

Geometry Post-Course Test (continued)

11. Which reason corresponds with the third statement in the proof, "△QRS ≅ △STQ?"

Ⓐ SAS Congruence Theorem

Ⓑ SSS Congruence Theorem

Ⓒ ASA Congruence Theorem

Ⓓ AAS Congruence Theorem

Given $\overline{QT} \cong \overline{SR}$, $\overline{QR} \cong \overline{ST}$
Prove △QRS ≅ △STQ

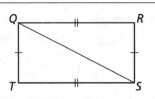

STATEMENTS	REASONS
1. $\overline{QT} \cong \overline{SR}$, $\overline{QR} \cong \overline{ST}$	1. Given
2. $\overline{QS} \cong \overline{QS}$	2. Reflexive Property of Segment Congruence
3. △QRS ≅ △STQ	3. _____

12. What is $m\angle WYZ$ in rhombus WXYZ?

Ⓐ 41°
Ⓑ 49°
Ⓒ 82°
Ⓓ 98°

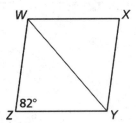

13. Complete the congruence statement.

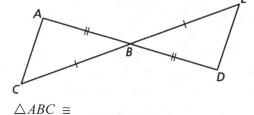

△ABC ≅ _____

14. Which of the following statements are incorrect? Select all that apply.

Ⓐ The centroid of a triangle is always inside the triangle.

Ⓑ The circumcenter of a triangle is equidistant from the vertices of the triangle.

Ⓒ The angle bisectors of a triangle intersect at the incenter.

Ⓓ The three medians of a triangle intersect at the circumcenter.

Ⓔ The orthocenter of a triangle is always inside the triangle.

15. Which property illustrates the statement "If $a = b$, then $b = a$?"

Ⓐ Symmetric Property of Equality

Ⓑ Reflexive Property of Equality

Ⓒ Transitive Property of Equality

Ⓓ Multiplication Property of Equality

16. What is the diameter of a circle with an area of 824.5 square inches?

Ⓐ about 8.1 in.

Ⓑ about 16.2 in.

Ⓒ about 28.7 in.

Ⓓ about 32.4 in.

Name _____ Date _____

Geometry Post-Course Test (continued)

17. A triangle, △ABC, is similar to △DEF. The longest side of △ABC is 45 units. What is the perimeter of △ABC?

Ⓐ 46 units
Ⓑ 58.5 units
Ⓒ 103.5 units
Ⓓ 207 units

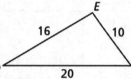

18. ∠ABC and ∠LMN are supplementary angles, $m\angle ABC = (3x - 15)°$, and $m\angle LMN = (5x + 9)°$. What is $m\angle ABC$?

Ⓐ 21°
Ⓑ 54.75°
Ⓒ 69°
Ⓓ 125.25°

19. What is the value of z?

Ⓐ 5
Ⓑ 6
Ⓒ 7
Ⓓ 8

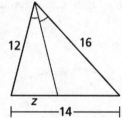

20. What is the measure of the exterior angle?

Ⓐ 85°
Ⓑ 95°
Ⓒ 122°
Ⓓ 153°

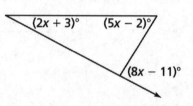

21. A manufacturer notices that 3% of bolts are defective. A sorting machine is 99% accurate at identifying correctly made bolts and 95% accurate at identifying defective bolts. What is the probability that a random bolt is correctly sorted by the machine? Write your answer as a decimal rounded to the nearest thousandth.

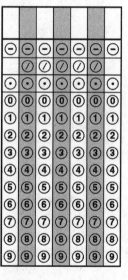

22. What is AB?

Ⓐ 4.6
Ⓑ 12.6
Ⓒ 14.9
Ⓓ 19.5

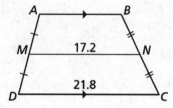

23. What is the height of the pyramid?

Ⓐ 1 ft
Ⓑ 3 ft
Ⓒ 6 ft
Ⓓ 9 ft

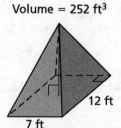

Volume = 252 ft³

Geometry Post-Course Test (continued)

24. What is the value of *y*?

Ⓐ 2
Ⓑ 3
Ⓒ 4
Ⓓ 8

25. What is *NP*?

Ⓐ 22
Ⓑ 25
Ⓒ 32
Ⓓ 37

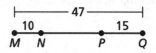

26. What can you conclude from the diagram?

Ⓐ $DE = GH$
Ⓑ $DE > GH$
Ⓒ $DE < GH$
Ⓓ No conclusion can be made.

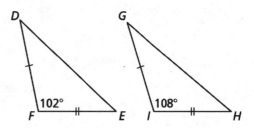

27. What is the value of *b*?

Ⓐ $2\sqrt{7}$
Ⓑ $4\sqrt{3}$
Ⓒ $2\sqrt{21}$
Ⓓ $4\sqrt{7}$

28. Which side lengths form an acute triangle?

Ⓐ 5, 12, 13
Ⓑ 8, 10, 14
Ⓒ 6, 7, 11
Ⓓ 7, 11, 12

29. What is the inverse of the conditional statement?
Conditional Statement: If a polygon is regular, then its sides are congruent.

Ⓐ If a polygon's sides are congruent, then it is regular.
Ⓑ If a polygon is not regular, then its sides are not congruent.
Ⓒ If a polygon's sides are not congruent, then it is not regular.
Ⓓ A polygon is regular if and only if its sides are congruent.

30. What is the slope-intercept form of the equation of the line passing through the point $(6, -1)$ that is parallel to the line $2x - 3y = 8$?

_____

Geometry Post-Course Test (continued)

31. The midpoint of $\overline{CD}$ is $M(-1, 7)$ and one endpoint of $\overline{CD}$ is $D(4, 5)$. What are the coordinates of C?

Ⓐ $(-6, 9)$

Ⓑ $(9, 3)$

Ⓒ $(-10, 4)$

Ⓓ $(1.5, 6)$

32. You flip two coins and randomly choose a letter from A to K. How many outcomes are possible?

Ⓐ 13

Ⓑ 15

Ⓒ 22

Ⓓ 44

33. What is an equation of the parabola?

_____

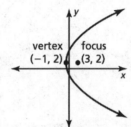

vertex (−1, 2)   focus (3, 2)

34. What is the distance from point A to $\overrightarrow{XZ}$? Round to the nearest hundredth.

_____ units

35. What is the radius of $\odot C$?

Ⓐ 2

Ⓑ 5

Ⓒ 12

Ⓓ 13

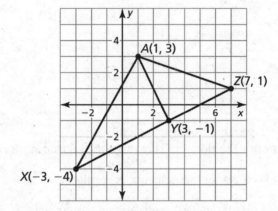

36. What is $m\angle F$?

Ⓐ about 34.1°

Ⓑ about 41.5°

Ⓒ about 75.6°

Ⓓ about 104.4°

Geometry Post-Course Test (continued)

37. ∠D is an acute angle and sin D = 0.719. What is m∠D? Round to the nearest tenth.

degrees

38. A quarter of a circle is removed from a square, as shown. What is the perimeter of the shaded region? Round to the nearest tenth.

cm

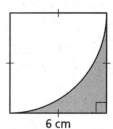

6 cm

39. On a recent test, students chose a method to use in order to study. After the test, the results were recorded in a two-way table. What is P(pass | study guide)?

Ⓐ about 17.6%

Ⓑ about 49.1%

Ⓒ about 82.4%

Ⓓ about 96.6%

		Grade	
		Pass	Fail
Method	Flashcards	10	4
	Study guide	28	6
	Reread notes	19	3

40. The endpoints of a diameter of a circle are (5, −2) and (11, 8). What is the standard equation of the circle?

41. What is MN?

Ⓐ 7

Ⓑ 13

Ⓒ 14

Ⓓ 26

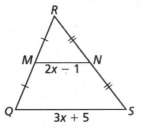

42. What is m∠Y?

Ⓐ 40°

Ⓑ 50°

Ⓒ 80°

Ⓓ 160°

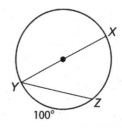

Geometry Post-Course Test (continued)

43. Which of the following properties and definitions are needed to complete the proof? Select all that apply.

- Ⓐ Transitive Property of Equality
- Ⓑ Subtraction Property of Equality
- Ⓒ Substitution Property of Equality
- Ⓓ Definition of right angle
- Ⓔ Symmetric Property of Equality
- Ⓕ Definition of angle bisector

Given ∠DBE is a right angle.
Prove ∠ABD and ∠CBE are complementary.

STATEMENTS	REASONS
1. ∠DBE is a right angle.	1. Given
2. m∠DBE = 90°	2. _____
3. m∠ABD + m∠DBE + m∠CBE = 180°	3. Angle Addition Postulate
4. m∠ABD + 90° + m∠CBE = 180°	4. _____
5. m∠ABD + m∠CBE = 90°	5. _____
6. ∠ABD and ∠CBE are complementary.	6. Definition of complementary angles

44. What is the sum of x and y?

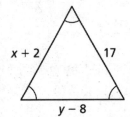

45. What is the sum of the measures of the interior angles of a convex 35-gon?

_____ degrees

46. Two sides of a triangle have lengths 3 centimeters and 10 centimeters. Write an inequality to represent the possible lengths, x (in centimeters), of the third side.

_____

Geometry Post-Course Test (continued)

47. What angle pair are ∠8 and ∠9?

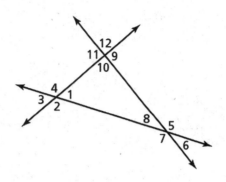

- Ⓐ corresponding angles
- Ⓑ alternative interior angles
- Ⓒ alternate exterior angles
- Ⓓ consecutive interior angles

48. What additional information do you need to prove that △WXY ≅ △WZY by the ASA Congruence Theorem?

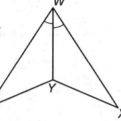

- Ⓐ ∠X ≅ ∠Z
- Ⓑ ∠WYX ≅ ∠WYZ
- Ⓒ $\overline{WX} \cong \overline{WZ}$
- Ⓓ $\overline{XY} \cong \overline{ZY}$

49. You make a five-digit password from the numbers 0–9. You cannot repeat any digit. How many passwords are possible?

- Ⓐ 252
- Ⓑ 2002
- Ⓒ 30,240
- Ⓓ 100,000

50. What is $m\angle 3$?

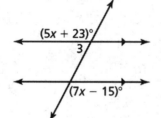

- Ⓐ 62°
- Ⓑ 85.3̄°
- Ⓒ 94.6̄°
- Ⓓ 118°

51. What is the surface area of the hemisphere?

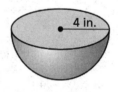

- Ⓐ about 100.5 in.²
- Ⓑ about 134.0 in.²
- Ⓒ about 150.8 in.²
- Ⓓ about 201.1 in.²

52. Three vertices of a parallelogram are $(-1, 5)$, $(3, 2)$, and $(1, 0)$. Which of the following could be the fourth vertex? Select all that apply.

- Ⓐ $(-3, 3)$
- Ⓑ $(5, -3)$
- Ⓒ $(-5, 1)$
- Ⓓ $(1, 7)$
- Ⓔ $(-7, 6)$

Name _____ Date _____

Geometry Post-Course Test (continued)

53. △ABC has vertices A(3, −5), B(1, −4), and C(4, −2). Which graph shows the image of △ABC after a reflection in the y-axis, followed by a rotation of 270° about the origin?

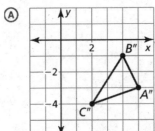

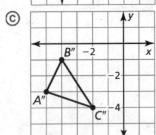

54. What is the surface area of the cone? Round to the nearest tenth.

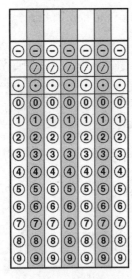

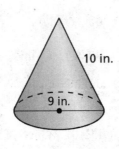

55. Point N is the incenter of △QRP, NY = 5x + 7, and NZ = 8x − 8. What is NX?

units

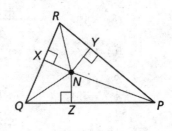

56. What is the volume of the cylinder?

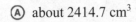

Ⓐ about 2414.7 cm³

Ⓑ about 6281.6 cm³

Ⓒ about 9076.3 cm³

Ⓓ about 36,305.0 cm³

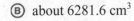

57. m∠XYZ = 130°. What is m∠WYX?

Ⓐ 20°

Ⓑ 56°

Ⓒ 65°

Ⓓ 74°

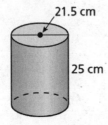

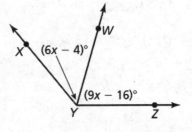

246 Geometry
Practice Workbook and Test Prep